MW00835576

BUSINESS FUNDAMENTALS FOR ENGINEERING MANAGERS

BUSINESS FUNDAMENTALS FOR ENGINEERING MANAGERS

C. M. CHANG

MOMENTUM PRESS, LLC, NEW YORK

Business Fundamentals for Engineering Managers
Copyright © Momentum Press®, LLC, 2014.

All rights reserved. No part of this publication may be reproduced, stored in a retrieval system, or transmitted in any form or by any means—electronic, mechanical, photocopy, recording, or any other—except for brief quotations, not to exceed 400 words, without the prior permission of the publisher.

First published by Momentum Press®, LLC
222 East 46th Street, New York, NY 10017
www.momentumpress.net

ISBN-13: 978-1-60650-478-9 (print)
ISBN-13: 978-1-60650-479-6 (e-book)

Momentum Press Engineering Management Collection

DOI: 10.5643/9781606504796

Cover design by Jonathan Pennell
Interior design by Exeter Premedia Services Private Ltd., Chennai, India

10 9 8 7 6 5 4 3 2 1

Printed in the United States of America

Dedicated to my loving family, wife Birdie Shiao-Ching, son Andrew Liang Ping, son Nelson Liang An, daughter-in-law Michele Ming Xiu, grandson Spencer Bo-Jun, and granddaughter Evya Bo-Ting

Abstract

Engineering managers and professionals make long and lasting impact in industry by regularly initiating and completing technology-based projects, as related to new product development, new service innovation or efficiency-centered process improvement, or both, to create strategic differentiation and operational excellence for their employers. They need certain business fundamentals that enable them to make decisions, based on both technology and business perspectives, leading to new or improved product/service offerings, which are technically feasible, economically viable, marketplace acceptable, and customer enlightening. Peter Drucker said, "Making good decisions is a crucial skill at every level."

This book consists of three sets of business fundamentals. The chapter "Cost Accounting and Control" discusses service and product costing, activity-based costing to define overhead expenses, and risk analysis and cost estimation under uncertainty. The chapter "Financial Accounting and Analysis" delineates the key financial statements, financial analyses, balanced scorecard, ratio analysis, and capital asset valuation, which includes operations, opportunities, and acquisition/mergers. The chapter "Marketing Management" reviews marketing functions, marketing forecasting, marketing segmentation, customers, and other factors affecting marketing in making value-adding contributions.

The new business vocabulary and useful analysis tools presented in this book will enable engineering managers to become more effective when interacting with senior management, and to ready themselves for assuming higher-level corporate responsibilities. It should be of great benefit to engineering managers and professionals who aspire to add value of increasing magnitude to their employers over time, based on their perspectives and decision-making capabilities, broadened by having acquired the useful business fundamentals. As all discussions of these topics are enhanced by examples, this book is particularly suitable for self-study, college courses, or as a desk reference by engineering managers and professionals. After reading this book, they will have the vocabulary, the broadened perspectives, and the tools to effectively apply the discussed business concepts and become increasingly effective leaders.

KEY WORDS

business fundamentals, business perspectives, cost accounting, financial accounting and analysis, management, marketing management

Contents

Preface		xi
Chapter 1	Introduction	1
Chapter 2	Cost Accounting and Control	7
Chapter 3	Financial Accounting and Analysis	75
Chapter 4	Marketing Management	141
Chapter 5	Conclusions	217
Notes		225
References		229
Index		237

Preface

A firm's efficacy of achieving business success is based on being able to provide a product/service better than that of its competitors, such as selling a lower price, offering richer features, functioning with higher reliability, being more customizable, and/or adding more value. For employees to make useful contributions in this respect, they need special know-how and outstanding preparation. This book covers the fundamentals of business management, which includes cost accounting, financial accounting and analysis, and marketing management. It is to enable engineering managers and professionals to facilitate their interactions with peer groups and senior managers and to make decisions based on both technological and business perspectives so that such decisions will achieve a broad-based acceptance by key decision makers and peers involved and receiving strong management support of its implementation.

Specifically, the contents of this book facilitate the decision making related to cost, finance, and services, as well as discounted cash flow and internal rate of return, especially during the product/service design phase in which a major portion of the final costs of products/services will be defined. Activity-based costing (ABC) is presented to define indirect costs related to products/services. It is useful for engineering managers and professionals to fully understand the costs (direct and indirect) of their products/services and strive to reduce them constantly.

By understanding the project evaluation criteria and the tools of financial analyses, engineering managers and professionals will be in a better position to secure project approvals by the senior leaders of the organization and to assess the health of a given external enterprise (such as suppliers or business partners to be), using the balanced scorecards. Also discussed will be economic value added (EVA), which determines the real profitability of the enterprise above and beyond its cost of capital deployed. For engineering managers and professionals to lead, a major challenge is the initiation, development, and implementation of major technology-based projects that are technically feasible and economically viable, resource conserving, and marketplace acceptable, while contributing to the long-term profitability of the company.

The important roles of marketing in any profit-seeking enterprise are self-evident. A critical step to developing technological projects that are attractive to the marketplace is the acquisition and incorporation of customer feedback. Engineering managers and professionals are expected to actively make contributions to support the marketing efforts. Many progressive enterprises are increasingly concentrating on customer relationship management to grow their businesses. Such a customer focus is expected to continue to serve as a key driving force for product/service design, project management, process improvement, productivity enhancement, customer services, and many other customer-centered activities.

Many illustrative examples are introduced to enhance the description of these business concepts. The use of business vocabulary and the associated business perspectives will greatly substantiate the decisions made by engineering managers and professionals.

I like to express my sincere appreciation to the State University of New York at Buffalo (SUNY-Buffalo) for the opportunities of having taught the graduate courses on Engineering Management there in the past 25 years. These courses included detailed discussions on cost accounting, financial accounting and analysis, and marketing management. During a part of this 25-year period, I was employed full time at Praxair, a Fortune-100 company specialized in industrial gases, to develop R&D technologies and conduct business-development activities. After I left Praxair and joined SUNY-Buffalo, I was involved in the development of a Master's Degree Program in Service Systems Engineering at its Department of Industrial and Systems Engineering in 2006, and was appointed as its full-time director for a brief period of time. Both sets of industrial and academic work experience offered me excellent opportunities to gain useful insights that have benefited me when writing this book.

It is my pleasure to thank Steward Mattson, senior vice president and COO of Business Expert Press/Momentum Press to have invited me to serve as the editor of a new Momentum Press Collection on Engineering Management, this book being the first of this series. I also want to acknowledge the able assistance of the Momentum Press, especially Shoshanna Goldberg, who made the publication process efficient and pleasant.

C. M. Chang, PhD, MBA, PE
State University of New York at Buffalo,
Buffalo, New York, USA

CHAPTER 1

INTRODUCTION

This book is written for engineering managers and professionals who are college graduates in a STEM (science, technology, engineering, and math) field and engage in one or more of the following types of work in industry: (a) design and develop new products and services; (b) manage projects of various kinds, such as operation of plants, installation of equipment, upgrade of facilities, and process improvement to cut cost and speed up cycle time, for which there are well-defined objectives, budgets, deadlines, and constraints to satisfy, involving different resources and talented people; (c) create actionable solutions to problems as related to technology deployment, operation, business partnership, supply chain development, and others; and (d) lead or participate in other activities (such as strategic planning, acquisition and merger assessment, technology introduction, supplier evaluation, and others) that require consideration from the standpoints of cost, finance, and marketing, in addition to that of technology and engineering.

Engineering managers and professionals need a set of business skills, which enable them to comfortably exercise judgments and make decisions from both the engineering and business perspectives. Having acquired skills in business fundamentals will benefit them in many ways: (a) **offer** decisions that are broad-based to better satisfy a diverse set of stakeholders, each having different demands and needs. This is a consequence of all businesses becoming increasingly complex rendering technology/engineering being just one of many important aspects to ponder. (b) **expand** own perspectives and view angles to benefit from seeing the world in broadened perspectives, without getting lost in the details. (c) **enhance** the prospective thinking while developing vision to create new opportunities, pinpoint new problems, recognize new patterns, detect new barriers, foray into a new market, war-game some future scenarios, and becoming experts who can see around the corners. (d) **burnish** own leadership credentials and participate in decision making regarding new corporate strategies to ferret out competitive advantages. (e) **protect** oneself from being too narrowly focused on technology/engineering aspects and be hurt unknowingly by forces outside of one's control. For example, by observing global trend, industrial changes, and enterprise level performance, some might decide to move to another career option, before it is too late. (f) **gain** new insights in business fundamentals and become well versed in assessing financial publications and other data sources, in order to pounce on investment opportunities (e.g., stock market, real estate, and foreign investment opportunities).

Coplin[1] suggested that professionals need to master 10 skills to succeed, which include: (a) taking responsibility, (b) developing physical skills, (c) communicating verbally, (d) communicating in writing, (e) working directly with people, (f) influencing people, (g) gathering information, (h) using quantitative tools, (i) asking and answering the right questions, and (j) solving problems. The general validity and usefulness of these skills are self-evident, as most of these skills should have been learned while in college or by self-study. However, to be effective in today's business environment, "asking and answering the right questions" require the use of pertinent vocabulary, domain concepts, knowledge, and insights. Only with the right preparation in business fundamentals will such actions of "asking and answering questions" by engineering managers and professionals be effective. The same goes with the skills related to "communications," "influencing people," and "solving problems."

Many for-profit organizations have the business mission to offer products or services to individual customers or corporate clients. For products or services to achieve sustainable profitability, they need to be (a) technically feasible to produce, (b) financially viable to design and generate, (c) strategically differentiable from competition, and (d) marketplace acceptable for enough customers to purchase. The acceptance in the marketplace requires that the novel products/services meet customers' current and future needs, sell at a price that is competitive with respect to other options customers have, produce enough gross margin for the company to generate profitability, distribute the products/services in ways convenient to the customers, and secure customer feedback to regularly update the offerings in order to retain many of them over time.

Engineering managers and professionals are generally talented in making contributions to assure technological feasibility of products and services, as many of them are well trained in STEM disciplines. Highly useful will be knowledge in cost accounting (including product costing, activity-based costing, target costing principle, and the application of

Monte-Carlo simulation method), which will enable them to assure that the products/services meet specific cost targets and remain economically viable, under a given pricing scenario. Understanding the basic tools in finance and accounting will allow them to readily check the financial viability of any technology project (such as the development of new products/services) using the net present value (NPV) analysis, which involves the use of a multiyear income statement. Exposure to marketing management concepts is useful, in order for them to become familiar with the tools available to reach customers by (a) understanding their current and future needs for the features of products/services in questions; (b) networking with them for possible collaboration for future product/service design; (c) becoming aware of customers' usage patterns to foster improved customization; (d) appreciating customer's buying decision, including pricing sensitivity; and (e) recognizing customers' preferences regarding service, maintenance, and repair. Knowledge is a decisive and competitive power to an enterprise, if it is created, maintained, and widely applied to add value.

The chapters on cost accounting, financial accounting, and marketing management in this book will help engineering managers and professionals to acquire these business fundamentals so that they are in a position to quickly and independently assess the technical, financial, and marketing attractiveness of any new product/service ideas they may come up with, thus, improving the effectiveness by which they could contribute to the business success of their employers.

For example, when suggesting a new product for the company to design, produce, and market in a given region, senior managers will want to know how novel the product concept is; how much will be the unit cost to produce; how likely will this product be successful at the gross margin level of say 35 percent, in view of the competition in the marketplace; what initial investment will it require; and what might be its projected NPV to the company. If these questions cannot be answered by the project team, the new product idea will quite likely not fly, even if its design concept is highly novel and readily patentable. The answers will need to be prepared by people who are well versed with the technologies involved as well as the basic business fundamentals discussed in this book.

Specifically, three chapters are elucidated in this book. *Chapter 2* addresses the issues related to cost accounting. The fundamentals of engineering economy are briefly reviewed in its Appendix, where a large number of examples are included to illustrate the basic principles and the use of these tools. Specific topics such as inventory accounting, depreciation accounting, product cost accounting, target costing, activity-based costing, and others are included. *Chapter 3* introduces financial accounting and analysis, and the important financial documents of income statement, balance sheet, and funds flow statement are introduced. The industrial standard of evaluating the financial viability of any technology project (e.g., developing a new product or service, retrofitting an old machinery, activating a productivity enhancing procedure, initiating a Lean Six Sigma project, etc.) requires the formation of a multiyear income statement, in which the cost of the project and its anticipated benefits are assessed to determine the project's NPV. Financial statements are the standard instruments to record the corporate performance. They are quite easy for engineering managers and professionals to master. *Chapter 4* covers marketing management. Without being able to market the company's products or services, no profitability can be achieved. Companies pay attention to 4Ps—(a) product (design, features, functionalities, value); (b) price (in view of competitive offerings); (c) promotion (channels, messages); and (d) placement (also called distribution that represents the ways to bring the products to customers). For marketing services, companies will add additional three more Ps—(e) process (the procedure of conducting activities, such as order processing, customer inquiries, problem solving, etc.); (f) people (friendliness and capabilities of customer-facing agents); and (g) physical evidence (e.g., color and design of lobby building, newness of office furniture and equipment, and dress codes of service agents). Companies are most interested in knowing more about the product features and service functionalities of value to potential customers, their buying criteria, and their anticipated future needs, as these inputs would help improve the design of future products and services.

The final chapter (*Chapter 5*) includes a few concluding remarks, reiterating the value of acquiring business fundamentals to engineering managers and professionals. References are cited for those who like to read on further in becoming well versed with additional details related to these business fundamentals.

There are a large number of colleges (such as MIT, Carnegie Mellon, University of Texas, Stanford, University of Pennsylvania, etc.) that offer MBA degrees with an Information technology concentration. In spring 2014, Cornell University at Ithaca plans to launch a one-year MBA degree program to give engineers a grounding in business management skills. Based on these educational examples, it should indeed be quite useful for technologists, besides IT engineers, to become well versed in some business fundamentals.

A number of themes permeate this book, which is written to facilitate the self-study of business fundamentals by engineering managers and professionals. It includes many examples with answers so that relevant concepts are clearly illustrated, as the saying goes: "Never doubt the power of examples." Currently, there are no books like this available in the marketplace, as described by the following short book reviews:

1. Cather et al.'s[2] book covers (a) organizing skills, (b) human resource management, (c) law, (d) project management, (e) money in the organization, (f) meeting customer's needs, (g) information technologies, and (h) electronic commerce. It is written for undergraduate students. The Cather book has a total of seven chapters on organizations and organizing, human resource management, law, project management, money in the organization, meeting customer needs, information technology, and electronic commerce. There is little overlap in coverage between the Cather book and this one of mine.
2. Brown's[3] book is written primarily for utility engineers. It covers the following chapters: (a) Utilities, (b) Accounting, (c) Economics, (d) Finance, (e) Risks, (f) Financial Ratios, (g) Rate Making, (h) Budgeting, and (i) Asset Management. Written specifically for utility engineers, Brown's book covers regulation, rate making, accounting, finance, risk management, economics, budgeting, and asset management. It does not cover marketing management, which is extremely valuable in the industrial world of importance to engineering managers.
3. McCubbrey's[4] book is written primarily for entrepreneurs and covers the following 15 chapters: (a) The Business Ecosystem, (b) The Mind of Entrepreneur, (c) Business Models and Marketing, (d) Organize and Lead an Entrepreneurial Venture, (e) Selecting and Managing Team, (f) Marketing on a Global Scale, (g) Operations Management, (h) Securing and Managing External Relations, (i) Financial and Managerial Accounting, (j) Leveraging with Information Technology, (k) Competitive Intelligence, (l) Business Ethics, (m) Adding Products and Services, (n) International Business for the Entrepreneur, and (o) Growth Strategy for Start-ups. Indeed, many of these topics are broadly useful for professionals working in industry, but they do not overlap with those covered in this book of mine.
4. Babson's[5] book is written for people with general background and has the following chapters: (a) The Use of Statistics, (b) Fundamentals and What They Foretell, (c) Making Figures Talk, (d) Forecasting Business Condition, (e) The Seesaw of Supply and Demand, (f) Scientific Purchasing, (g) Managing Men and Economic Law, (h) Solving the Production Problem, (i) Methods of Marketing,

(j) Selling a City, (k) Selling an Industry, (l) The Trend of Business, (m) Financial Independence, (n) Investing Your Income, (o) Successful Speculation, (p) Business Problems, (q) Investment Problems, (r) A Continuous Working Plan for Your Money, and (s) Conclusions. Again, its coverage is broad and there is little overlap with this book of mine.

Alexander Graham Bell said: "Before anything else, preparation is the key to success." The business fundamentals discussed in this book will help prepare engineering professionals to become better contributors, by thinking more broadly when making decisions and communicating in suitable business language (as related to cost, profit, gross margin, market segmentation, customer feedback, market share, and others) to garner support from others in different functional divisions. Doing so will significantly expand the likelihood that their judgments and decisions would be readily accepted by others. As the Japanese proverb says: "None of us are as smart as all of us," and gaining others' acceptance and understanding is of great importance in achieving success in industry.

Strong timber does not live at ease, the stronger the breeze, the stronger the trees. Demonstrating such decision and communication capabilities is essential for any engineering manager and professional, who brings his or her deep knowledge, experience, and wisdom to bear to enhance the enterprise's capability of shaping outcome and ameliorating its fortune.

CHAPTER 2

Cost Accounting and Control

2.1 INTRODUCTION

Cost estimation is the assessment of the value of resources consumed in the generation of a product/service offered for sales. Cost accounting and control are very important management functions in both profit-seeking and nonprofit organizations.

A profit-seeking enterprise strives to maximize its financial gains (e.g., sales revenue minus costs) for its investors. These gains can be sustained over time only if all stakeholders of the firm (e.g., investors, customers, employees, suppliers, business partners, and the community in which the firm operates) are reasonably satisfied. A nonprofit organization (e.g., the United Way, the Ford Foundation, government agencies, educational institutions, church organizations, etc.) seeks to maximize the organization's impact on its respective service recipients and target audience while minimizing operation costs.

This chapter covers the basics of cost accounting. The discussions focus on the costing of services and products.[1] After some commonly utilized accounting terms are introduced, the costing of products follows, including the estimation of direct costs absorbed into the company's inventory. The complex problem of assigning indirect costs to products is illustrated by the conventional method of using overhead rates, as well as by the more sophisticated method of activity-based costing (ABC).

Estimation of costs with uncertainties is then presented. The Monte Carlo simulation is introduced as an effective method to account for cost uncertainties. Its superiority over the conventional estimation method, which uses deterministic data, is demonstrated through examples of the output distribution functions of the Monte Carlo simulation. Finally, inventory accounting is addressed to arrive at the all-important cost of goods sold (CGS).

The cost analysis of a single period versus multiple periods is elucidated in the Appendix, including topics such as the time value of money and compound interest formulas, and others related to engineering economy. It is important for all engineering managers and professionals to become well-versed in cost accounting, as part of their requisite skills to succeed, as they need to know how to estimate service/product costs, manage overhead costs, initiate steps to further reduce them, and bestow their employer's business with predictable cost advantages.

2.2 BASIC TERMS IN COST ACCOUNTING

Engineering managers need to become familiar with the standard vocabulary used by cost accountants or cost engineers, as costs are important bases for evaluating corporate performance, conducting profitability analysis of projects, and making managerial decisions. While the cost-accounting systems used by various firms do not need to strictly follow the Generally Accepted Accounting Principles (GAAP) adopted by the financial accounting profession, engineering managers are still advised to understand the meaning of various accounting terms in order to ensure that their cost-based decisions are meaningful and understandable. The following is a general set of accounting terms used by many firms:[2]

1. **Cost center:** An organizational unit responsible for controlling costs related to its functional objectives (e.g., R&D, procurement, operations, engineering, design, and marketing).
2. **Inventory costs:** The total sum of product costs, which are composed of the direct costs and indirect costs related to the manufacturing of the products currently stored in warehouses.
3. **Direct costs:** Material and labor costs associated with the manufacturing of finished product/service.
4. **Indirect costs:** All overhead costs (e.g., rent, procurement, depreciation, supervision, supplies, power, quality control, safety, and others) indirectly associated with the fabrication of products/services involved.
5. **Fixed costs:** Costs that do not strictly vary with the volume of products involved, such as the general manager's salary, rent for the facility, machine depreciation charges, and local taxes.
6. **Variable costs:** Costs that vary in direct proportion to the volume of products involved, including, for example, material, labor, and utilities.

7. **Step function costs:** Costs that would experience a step change when a specific volume range is exceeded; for example, the factory rent that may change stepwise if new floor space must be added because of the increased production volume.
8. **Contribution margin:** The product price minus unit variable cost, which is the economic value contributed by selling one unit of the product to defray the fixed cost already committed for the current production facility.
9. **Cost pool:** An organizational unit wherein costs incurred by its activities performed for specific products (or other cost targets) are accumulated for subsequent assignments.
10. **Cost drivers:** Bases used to allocate indirect costs to products. Products drive the consumption of resources and the utilization of resources incurs costs. Examples of cost drivers include floor space, head counts, number of transactions, number of employees, labor hours, machine hours, number of setups, material weight, and others.
11. **Cost objects:** Targets to allocate indirect costs, such as products and services sold by the firm, and customer groups served.
12. **Budget:** A quantitative expression in dollar value of a project or a plan of action. Examples include production budget, product design budget, engineering budget, R&D budget, sales budget, marketing budget, and advertising budget. Typically, budgets span a specific period of time (e.g., a month, a quarter, or a year).
13. **Standard costs:** Direct and indirect costs budgeted for products. The standard costs are defined by using estimations or historical costs.
14. **Variance:** The difference between standard costs and actual costs. Such variance could be the result of price variation, quantity change, technology advancement, and other factors. Conventionally, actual quantities are used when computing price variation to easily assess the procurement performance. On the other hand, the quantity-based variance is computed by using standard costs for an easy assessment of the production performance.
15. **Current costs:** Costs for the total efforts (e.g., physical efforts, raw materials, and service fees) that must be spent in order to carry out an activity or implement a plan. Current costs are typically used to inform managerial decision making.
16. **Opportunity costs:** The benefit of the second-best alternative that must be forgone because of a choice made for the first alternative. For example, an engineering manager who quits a job paying $100,000 a year to pursue a three-semester MBA degree at a

university incurs an opportunity cost at graduation of $150,000 plus an out-of-pocket cost of $90,000 for tuition and other fees. Opportunity costs are included in managerial decision making, but are not included in a cost-accounting system.

17. **Sunk costs:** Costs that have already been spent or incurred. Such costs are typically included in all cost-accounting systems, but they are not considered in any management decision making for the future.

These accounting terms are relatively easy for anyone to acquire and make use of.

2.3 PRODUCT AND SERVICE COSTING

Product costing and service costing, being one of the key responsibilities of engineering managers, are similar in that both encompass the direct costs (e.g., raw materials, labor) and indirect costs (e.g., general supports, maintenance, overhead, and others), which are incurred during the phases of generating the service or product involved. Direct costs are those that vary directly with the volume/quantity of the service or product produced; these are relatively easy to properly account for. Indirect costs, on the other hand, are somewhat difficult to allocate because of the complex and varied nature of these costs and their nonobvious relationships to the cost objects at hand. Cost objects are the targets (e.g., services, products and customers) for which costing is to be performed.

The traditional practice of general ledger costing involves estimating all overhead costs for the upcoming year in a single cost pool (e.g., factory overhead [FO], utilities, safety programs, training, salaries of foremen and compensations of factory managers). This total is then divided by the estimated number of labor hours to be worked. The result is an hourly overhead rate. For each new product/service, the required labor hours are first estimated. The total overhead cost for this new product/service is then equal to the respective labor hours required to produce it multiplied by the hourly overhead rate.

According to Wiese,[3] traditional costing systems accumulate costs into facility-wide or departmental cost pools. The costs in each cost pool are heterogeneous—of many different processes—and are generally not caused by a single method of resource utilization. Such a system allocates costs to products/services based on volume (e.g., units, direct labor [DL] hours, machine hours, or revenue dollars). The resulting unit costs for products or services may be overestimated or underestimated in this manner; specifically there will likely be a cost distortion of overcharging overhead to high-volume products and charging too little to low-volume products. A better method of allocating indirect costs is the ABC, which is introduced in the next section.

2.3.1 ACTIVITY-BASED COSTING

ABC is a cost-accounting technique by which indirect and administrative support costs are traced to activities and processes and then to the cost objects (e.g., services, products, or customers). It is based on the rationale that resources generate costs, activities utilize resources, and cost objects (e.g., products, services, and customers) consume activities.

ABC is built on the notion that an organization has to perform certain activities in order to generate products and services. These activities cost money. The cost of each of these activities is only measured by and assigned to those products or services requiring identifiable activities and using appropriate assignment bases (called cost drivers). The results of ABC analyses offer an accurate picture of the real cost of each product or service, including the cost of serving customers. Nonactivity costs (such as direct materials [DMs], DL, or direct outside services) do not need to be included because these costs are readily attributable to the specific product or service under consideration. ABC is most useful for companies with diverse products, service centers, channels, and customers, and for those companies whose overhead costs represent a large percentage of their overall costs for the product and service.[4]

In the service economy, direct manufacturing labor is no longer the overriding factor of production and the distinction between production and service departments has become decidedly blurred. The overall costs of products/services are more influenced by research, materials handling, procurement, equipment maintenance, quality control, and customer service requirements than by DL. The ABC technique accumulates costs into activity cost pools, in accordance with the groups of major activities or business processes. The costs in each cost pool are largely incurred uniformly by a single factor—the cost driver. ABC systems allocate costs to specific cost objects (products, services, customers, etc.) from the cost pools using these applicable cost drivers as the allocation bases. As a consequence, the cost information so provided is more accurate.[5]

According to Atkinson,[6] ABC is particularly useful to service companies because: (a) Most costs are indirect and appear to be fixed. Variable costs tend to be small and frequently near zero. (b) Most costs are capacity-related costs. These costs are based on the amount required, rather than the amount used. (c) Most costs are customer specific rather that customer independent. For service companies, it is highly desirable to define the differential profitability of individual customers—each may demand different amount of resources to serve—by applying ABC.

All engineering managers and professionals should learn to practice ABC, because the traditional method of allocating overhead uses only high-level information about costs, and the general ledger system does not provide information related to time and resources spent on assignments and activities. In contrast, a well-practiced ABC technique offers specific insights that include (a) a clearer picture for management as to which product or service generates profits and losses for the company, (b) the ability to track operating profits for specific cost objects (such as customers, orders, and products), (c) the ability to determine whether a service center is efficient or deficient, and (d) the possibility of exposing non–value-adding activities, which the company could reduce or eliminate to improve its operational effectiveness.

Even a company with an overall profitability may lose money on certain products, orders, and customers in the absence of detailed costing information created by ABC. According to the published best practices of some industrial pioneers (such as Honeywell Inc. and Coca-Cola®) on the use of ABC, simpler ABC models deliver better results.

ABC has become increasingly popular with industrial companies, partly because it is useful for organizations of any size and does not require a massive effort to implement, and partly because of increased processing capabilities of personal computers (PCs), reduced prices of ABC software products, and increased competition forcing companies to achieve a better understanding of their own product/service costs. There are several ABC software products in the market. Examples include ABM by SAS, HPCM by Oracle, RapidABC by Virtual Profit Solutions, and PCM by SAP.

2.3.2 SURVEY OF ABC USE IN COMPANIES

In July 2005, SAS conducted a survey of ABC uses among 529 companies in industries, of which 56 percent in services, 24 percent in manufacturing, and 20 percent in others. Forty-two percent of these companies had sales revenues at or below $100 million, and 26 percent were above $500 million.

The overall results showed that 35 percent of these companies were using ABC, 20 percent were engaged in piloting ABC, and another 32 percent were considering ABC. Only 10 percent of these companies had never considered ABC and 2 percent were no longer using ABC. The use of ABC increased with company size. ABC is being used or being considered for use by 71 percent of large companies, 58 percent of midsize companies, and 42 percent of small business. Within the service industry, 46 percent of financial services and 58 percent of communications companies use ABC. The primary use of ABC by all industries is for costing and cost control. About 80 percent of all ABC programs were initiated by people in the finance department.

This survey showed that ABC is important to all enterprises and useful for future technical professionals to know and master.

2.3.3 STEPS TO IMPLEMENT ABC

It is generally advisable to form a cross-functional team when implementing the ABC technique of allocating indirect costs. The team should determine the cost objects. Examples of cost objects include costs to serve customer; costs to purchase, carry, and process products; costs to order, receive, sell, and deliver products; and costs to perform other activities. The team then needs to define activities that represent homogeneous groups of work (such as accounting, machining, forging, and design) that lead to the cost objects.

Next to be determined are cost drivers. These are the agents that cause costs to be incurred in the activities. Cost drivers are factors that directly impact the cost of a given cost object. Examples of cost drivers are shown in Table 2.1.

Figure 2.1 shows a generic block-flow diagram for implementing ABC. It contains 10 major steps.

Table 2.1. Cost drivers

Activity	Cost driver
Loading	Tons
Driving	Miles
Invoice processing	Number of invoice
Machining	Machining hours
Material movement	Weight
Production	Number of products

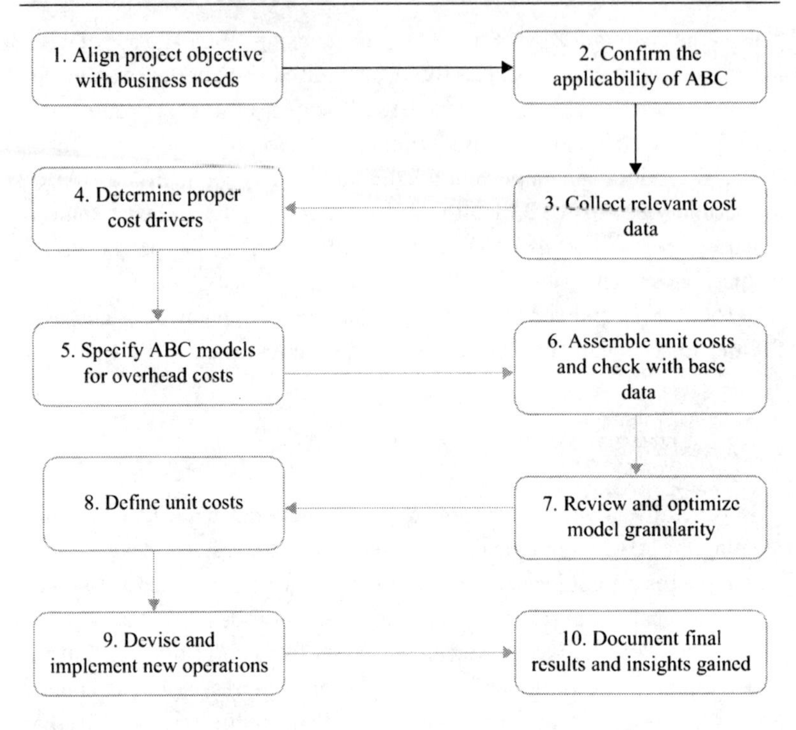

Figure 2.1. Block-flow diagram for Implementing ABC.

The details of these 10 steps are further described as follows:

1. The first very important step is to define an ABC project, whose objective aligns fully with business needs. The value of an ABC project, which would provide more structural knowledge regarding the cost objectives involved, must be convincing to leaders in the organization. Otherwise, corporate commitment, management priority, and availability of resources will likely be in doubt. ABC is best to offer such cost structural information, when large overhead costs are shared among a large number of cost objectives (such as products, services, or customer groups).
2. Make sure that ABC is indeed the right method to accomplish the defined project objective, based on the availability of cost data, manpower, and applicable techniques.
3. Collect the relevant data from sources such as general ledger, time sheets, procurement records, and other data sources, particularly those related to activities that add value to the cost objectives. Interview with applicable managers may be needed to help identify additional details that may be useful in categorizing the cost data.

4. Define cost drivers for the various resource-consuming activities. Sources for this information could be industrial best practices, engineering literature, and /or rational and logical assumptions.
5. Create an ABC cost model for allocating overhead costs to the specific cost objectives at hand, including all value-adding and resource-consuming activities.
6. The unit cost of the specific cost objective is then assembled. This result needs to be double checked to make sure that the total overhead cost agrees with the base data contained in the general ledger. If there are deviations, the ABC cost model must be readjusted to eliminate the differences.
7. The model is further reviewed to see if the ABC model's granularity is good enough for the purpose at hand. Past experience suggests that simple ABC models should be used in the beginning in order to quickly produce results and demonstrate value. Detailed ABC models with a high level of sophistication should be used only if outcome of simplified models justifies the additional efforts so required.
8. Define the final unit costs and determine the new insights gained from the ABC results.
9. Devise new strategies to take advantage of the ABC results. Make sure that these strategies are implemented effectively to realize added profitability. Examples of such new strategies may include promoting (a) promoting one product/service more than the other because of differences in gross margins; (b) changing the customer support service strategy in favor of customers who are more valuable than others; and (c) modifying pricing strategies because of the cost differences between products/services, and others.
10. Document results and preserve insights gained so that lessons are shared with others in the future.

2.3.4 *PRACTICAL TIPS FOR APPLYING ABC*

When initiating the process of creating an ABC system, it is highly recommended for the company to start with a small group (pilot group) of well-informed and cross-functional workers. The team should interview other workers about what they do in their jobs. The team members should be cognizant of the potential fears of job restructuring that some employees may have as a result of the ABC studies.

The team should start with the "worst" department so that immediate success may be used to get faster "buy-in" from top management. The key for ABC success is to use "close-enough" data. The team should keep the level of information manageable by avoiding being bogged down with minute details. On the other hand, an ABC system that is too broad and general will not be useful. The team may have to try out ABC cost models of different granularities on small scales to reach such "sweet spots." For companies attempting to employ ABC cost models for the first time, useful outputs can generally be expected in 6–12 months.

To be successful in implementing ABC techniques, company management needs to (a) define linkage of ABC to company's competitive strategies, (b) allocate sufficient financial resources (manpower, ABC software, and external consultants, if deemed needed), (c) conduct staff training in skills needed to collect data and design the ABC system, (d) secure clarity for the objectives of ABC systems, (e) establish proper evaluation and compensation policies, and (f) devote adequate management attention to gain useful ABC outcome.

It is important to note that the application of ABC techniques will not, by itself, reduce product/service cost or increase corporate profitability, as this technique only redistributes the overhead or indirect cost on an increasingly rational basis. However, it is the insight created by the ABC outcome that would be the basis for the company to initiate new or modified pricing, distribution, or other operations strategies, and the implementation of such strategies could produce value in the form of improved overall company profitability.

2.4 APPLICATIONS OF ABC IN INDUSTRY

2.4.1 ABC IN MANUFACTURING

Product costing in a product-centered company requires the computation of costs related to DM, DL, and FO for the following three operations:

- Raw materials (Stores)
- Work in progress (WIP)
- Finished goods (FG)

In computing the costs of goods sold (CGS), the inclusion of materials and labor costs is rather straightforward, as these direct costs are quite easy to track. The difficulty in product costing is the inclusion of indirect costs, namely the FO. For manufacturing operation in which the FO represents a large (e.g., 30 percent or higher) fraction of the total product cost, or the operation produces multiple products, or both, the precision with which to allocate the indirect costs is critically important. The following is a specific example of applying ABC to XYZ Manufacturing Company.

Let us assume that XYZ is a small manufacturing company with $10 million in annual sales. It makes components for the automotive industry, and the key processes involved are forging and machining. The product-related operating activities are as follows:

1. *Buying* steel bars from outside vendors.
2. *Testing* steel bars upon delivery and moving them into storage.
3. *Sending* the bars to the forging area when needed for an order, the point at which they are sandblasted and cut to desired lengths. Since most of the bars are large, they are then moved in bins that hold 20–25 pieces.
4. *Sizing* the bars and starting a forging operation where they are shaped. The bars are then moved to the in-process storage. In some cases, a steel bar may need to be forged up to three times.
5. *Transferring* the bars, for each forging procedure from in-process storage to the forging areas and then back to the in-process storage.
6. *Moving* the steel bars after the final forging from the in-process storage to the machining area where they are finished. The bars are then sent to finished-goods storage.
7. *Sorting*, packing, and loading the bars are done in the shipping area and onto trucks for delivery to customers.

Before using ABC, the company applied the traditional costing method that included the following steps:

1. Assign manufacturing costs to products by using a plant-wide costing rate on the basis of DL. The setup costs are included in the manufacturing overhead.
2. Determine the nonmanufacturing costs to products via a general and administrative (G&A) rate that is calculated as a percentage of the total cost.
3. Define the DL rate and the G&A rate on the basis of the actual results obtained for the preceding year.

The deficiencies of the traditional method are obvious. The traditional method is used because management does not know any better methods.

When implementing ABC, the company did not buy any ABC-specific software. Instead, it used a standard Excel spreadsheet program. Specifically, the company considered the following:

1. **Setup costs.** Management assigned equipment setup costs only to the steel bars in a given equipment process.
2. **Forging costs.** Depending on the weight of the steel bar involved, one or two operators may operate the forging press. Prior to forging, each steel bar must be induction heated, with the heating cost being dependent on the mass of the bar involved. Thus, the forging cost consists of three parts:
   (a) Press-operating costs on the basis of press hours
   (b) Production labor costs on the basis of labor hours
   (c) Induction-heating costs on the basis of the steel bar's weight
3. **Machining costs.** The machining centers do not require full-time operators. Once the machines are set up, workers load and unload parts for multiple centers. On average, 1 machine-worker hour is required for every 2½ machine hours. Thus, the costs of machine-shop workers are treated as the indirect costs assigned to products on the basis of machine hours.
4. **Material movement costs.** Depending on the size of the bar, the bin size, and the required forging and machining steps, the material movement cost could vary significantly from one bar to another. Thus, the material movement cost is assigned to each bar on the cost-per-move basis.
5. **Raw material procurement** and **order processing costs**. These are readily traceable on the basis of records on hand.

The ABC cost model for the XYZ Manufacturing Company is illustrated in Table 2.2. The final results of ABC implementation are impressive. The company's sales tripled and its profit increased fivefold after having made use of the new insights gained from the ABC outcome. Specifically, much of this improvement came from a more profitable mix of contracts generated by a pricing and quoting process that more closely reflects the actual cost structure of the company. Particularly useful are the isolation and measurement of material movement costs that result in operational changes for increased efficiency.

Example 2.1

A company makes and sells three technology products: A, B, and C. It has a production plant with 17,000 square feet of floor area, consisting

Table 2.2. ABC model for a manufacturing company

Cost categories	Forging press hour cost	Machine hour cost	Induction-heating cost	Material movement cost
Directly attributable costs	Depreciation	Depreciation	Depreciation	Depreciation
	Utilities	Utilities	Utilities	Utilities
	Manuf. supplies	Manuf. supplies	Manuf. supplies	Manuf. supplies
	Outside repairs	Outside repairs	Outside repairs	Outside repairs
		Straight-line wages		Straight-line wages
		Fringe benefits		Fringe benefits
		Payroll taxes		Payroll taxes
		Overtime premium		Overtime premium
		Shift premium		Equipment leases
Distributions	Maintenance	Maintenance	Maintenance	Maintenance
	Buildings and grounds	Buildings and grounds	Buildings and grounds	Buildings and grounds
	Manuf. engineering	Manuf. engineering	Manuf. engineering	Human resources
	Commodity overhead	Commodity overhead	Commodity overhead	Supervision
		Supervision		
Total	Total costs	Total costs	Total costs	Total costs
Rate	$ per press hour	$ per machine hour	$ per heating weight	$ per move

Table 2.3. Production hours for three products

Production-related entries	A	B	C
Machine setup (hours)	2	3	4
Machine operation (hours)	16	12	8
Assembly (hours)	4	3	2
Inspection/packing/shipment (hours)	2	2	2
Raw materials/unit of product (dollars)	950	430	640
Purchased components/unit of product (dollars)	100	80	90
Outsourced service/unit of product (dollars)	20	30	40
Number of units produced per year	700	900	550

of machine setup (2,000 square feet), machining operation (9,000 square feet), assembly (4,000 square feet), and inspection, packaging, and shipping activities (2,000 square feet).

The total annual expenditure for the plant is $200,000 for depreciation, $700,000 for utilities, $20,000 for phone and travel services, $150,000 for manufacturing supports, $200,000 for procurement, and $150,000 for supervision.

The labor hours and material costs required to manufacture the products are shown in Table 2.3. The labor charges are $25 per hour for machine setup; $35 per hour for machining operation; $30 per hour for assembly; and $20 per hour for inspection, packing, and shipping.

The company plans to sell product A at $5,000 per unit, product B at $4,500 per unit, and product C at $4,100 per unit. All products manufactured during the year are assumed to be sold successfully. Apply the ABC technique to determine the product cost and applicable gross margin for each product.

Answer 2.1

Let us first summarize the available operational data of the company, see Table 2.4.

Table 2.5 contains a detailed ABC cost analysis based on these data.

Table 2.4. Operational data

	Product A	Product B	Product C	Labor rate/hour ($)
Machine setup (hours)	2	3	4	25
Machine operation (hours)	16	12	8	35
Assembly (hours)	4	3	2	20
Inspection/pack/ship (hours)	2	2	2	
Raw materials/unit	$950	$430	$640	
Purchased components/unit	$100	$80	$90	
Outsourced services/ unit	$20	$30	$40	
Number of products made/year	700	900	550	
Total labor hours per unit	24	20	16	
Total material cost per unit	$1,700	$540	$770	

2.4.2 ABC IN BANKING AND FINANCIAL SERVICES

Buckeye National Bank[7] serves both retail and business customers. The services include paying checks, providing teller services, and responding to customers' service calls. All of these services represent labor-intensive activities. The resources involved are employees (salary and benefits), part-time workers, and those related to the operation of service call centers. The bank's traditional costing system suggests that it is more profitable for the bank to pursue additional retail customers and that business customers bring losses. As a result, the bank's retail customers grow, while the business customers remain stable. Yet the bank's profit is unexpectedly trending downward. Bank management becomes puzzled as to the reasons why.

A large number of cost data were made available from the banks' traditional accounting system. Tables 2.6 and 2.7 display these data.

Table 2.5. ABC analysis of Example 2.1

Coat items	Area	Percentage	Depreciation fraction	Product A	Product B	Product C	Basis of cost allocation
1. Depreciation: $200,000							
Setup	2,000	11.76	23,529.41	7.47	11.2	14.94	Setup Hours
Operation	9,000	52.94	105,882.35	64.17	48.13	32.09	Op. Hours
Assembly	4,000	23.53	47,058.82	28.52	21.39	14.26	Assem. hours
Inspection	2,000	11.76	23,529.41	10.94	10.94	10.94	Inspec. hours
Total	17,000		$200,000	111.11	91.67	72.23	
2. Utilities: $700,000							
Utilities per unit				424.24	318.18	212.12	Op. hours
3. Labor							
Labor cost per unit				$770	$625	$480	Labor hours
4. Manufacturing support: $150,000							
Manufacturing support per unit				82.57	68.81	55.05	Prod. hours/unit
5. Supervision: $150,000							
Supervision per unit				82.57	68.81	55.05	Prod. hours/unit

6. Procurement: $200,000					
Procurement per unit		$129.03	$65.12	$92.85	Mat. cost/unit
7. Phone and travel: $20,000					
Phone and travel per unit		9.3	9.3	9.3	
8. Summary of unit product cost					
Unit product cost	Raw materials	$950	$430	$640	
	Purchased components	$100	$80	$90	
	Outsourced service	$20	$30	$40	
	Depreciation	111.11	91.67	72.23	
	Utilities	424.24	318.18	212.12	
	Labor cost	$770	$625	$480	
	Manufacturing support	82.57	68.81	55.05	
	Supervision	82.57	68.81	55.05	
	Procurement	$129.03	$65.12	$92.85	
	Phone + travel	9.3	9.3	9.3	
	Total product cost/unit	**$2,679**	**$1,787**	**$1,747**	

(Continued)

Table 2.5. (*Continued*)

Coat items	Area	Percentage	Depreciation fraction	Product A	Product B	Product C	Basis of cost allocation
9. Gross margin	Price/unit			\$5,000	\$4,500	\$4,100	
	Gross margin			\$2,321	\$2,713	\$2,353	
	Gross margin %			46.42%	60.29%	57.40%	
	Total gross margin		**\$5,361,000**				

Note: Strategic decision may be made in favor of Product B, which has a high gross margin percentage. Note the comments:

(1) Total overhead cost is \$1,420,000 (= 200,000 + 700,000 + 20,000 + 15,000 + 200,000 + 150,000)
(2) Total material cost is \$1,650,000 (=700 (950 + 100 + 20) + 900 (430 + 80 + 30) + 550 (640 + 90 + 40))
(3) Total labor cost is \$1,365,500 (=700(2 × 25 + 16 × 35 + 4 × 30 + 2 × 20) + 900 (3 × 25 + 12 × 35 + 3 × 30 + 2 × 20) + 550(4 × 25 + 8 × 35 + 2 × 30 + 2 × 20))
(4) Sum of all costs is \$4,444,400

ABS is to redistribute the overhead costs to various products. After redistribution, the total cost should be exactly equal to \$4,444,400, no more and no less.

The product costs for A, B, and C are \$2,679, \$1,787, and \$1,747, respectively. The individual gross margins for A, B, and C are \$2,321, \$2,713, and \$2,553, respectively.

Table 2.6. General ledger cost data of Buckeye National Bank

Buckeye National Bank	Cost ($)
Salaries of check-processing personnel	700,000
Depreciation of equipment used in check processing	440,000
Teller salaries	1,000,000
Depreciation of equipment used in teller operations	200,000
Salaries of call center personnel	450,000
Tool-free phone line plus depreciation of related equipment	60,000
Total costs	2,850,000
Total profit	650,000

Table 2.7. Additional cost data of Buckeye National Bank

Service-based entries	Retail	Business
$ Value of check processed	$9,500,000	$85,500,000
Checks processed	570,000	2,280,000
Teller transactions	160,000	40,000
Number of customer calls	95,000	5,000
Annual profit (interests) per account	$10	$40

Table 2.8. Traditional cost solutions

Traditional system	Retail ($)	Business ($)
$ Value of check processed	9,500,000	85,500,000
Cost per $ processed	0.03	$0.03
Total cost	285,000	2,565,000
Cost per account	1.90	51.30
Annual Profit per account	8.10	–11.30

The solution obtained based on the traditional costing system is shown in Table 2.8. It indicates that the annual profit of the retail account is $8.10 whereas that of the business account shows a loss of $11.30. The logic of pursuing more analyses regarding retail customers appears to be compelling.

Table 2.9. Unit costs of three principal activities

Cost per check processed	\$0.40 = (700,000+440,000)/2,850,000
Cost for teller transition	\$6.00 = (1,000,000+200,000)/200,000
Customer inquiries	\$5.10 = (450,000+60,000)/100,000

Table 2.10. ABC solution

ABC cost assignment	Retail	Business
Paying checks (0.4 × 570K)	\$228,000	912,000
Teller transactions (6 × 160)	960,000	240,000
Call centers (5.10 × 95,000)	484,500	25,500
Total	1,672,500	1,177,500
Per account cost	\$11.15	\$23.55
Net profit per account	−1.15	16.45

An ABC pilot study was then initiated to define (a) percent of time each employee spends on the afore-mentioned three activities, and (b) costs associated with toll-free phone lines, depreciation of equipment used for paying checks, and providing teller services. The cost drivers were identified to be (a) number of checks processed for the check paying activity, (b) number of teller transactions for the teller service activities, and (c) number of calls received for responding to customer inquiries.

Table 2.9 summarizes the unit cost of these three activities. Table 2.10 shows the cost assignments and the per account cost. In fact, the retail accounts are shown to lose money, indicating that the bank policy of pursuing retail customers was based on an erroneous analysis of costing data.

Based on the ABC outcome, Buckeye National Bank was able to improve profitability by soliciting more business customers, while stop promoting the business of retail customers.

2.4.3 ABC IN HEALTH CARE

The traditional cost-accounting system tends to overestimate the unit cost of high-volume services and underestimate that of low-volume services.

When the indirect costs are large, often the case in health care, the cost of services may be seriously misrepresented by the traditional cost-accounting system.[8]

The MaxSalud Institute for High Quality Healthcare in Chicalyo, Peru, is a private, nonprofit organization funded by United States Agency for International Development (USAID) to provide health services to a low- and mid-income population of about 20,000 through two clinics and a central management support unit. This ABC application included: (a) the description of all departments, services, and their activities; (b) staff estimates of time spent on each activity and unproductive time; (c) estimated cost of all activities by each department using wage and other data; (d) records of activities and costs within and across departments to services provided; and (e) estimated service volumes from records to determine unit costs (cost/volume).

This ABC study identified 107 distinct activities at MaxSalud, including training and meetings. ABC derived unit costs that were generally higher than prior estimates and much higher than fees charged. Among others, the ABC study discovered that the primary activity of delivering a baby accounts for only 23 percent of the total unit cost, while 42 percent from secondary activities (i.e., admission, general services, and others), and 35 percent from overhead (i.e., the central management support unit). The study also revealed information on activities associated with unproductive times, such as repeating lab tests.

The study concluded that ABC is potentially very valuable to MaxSalud to set policy, manage expenditure, and even raise funds, but it requires reliable data systems for costs and service statistics, management attention, staff support, and technical assistance.

Several useful guidelines were suggested when applying ABC in developing countries:

1. ABC requires complementary accounting systems that provide reasonably accurate costs organized by cost category and department.
2. ABC requires accurate information on the volume of services provided.
3. Access to and strong cooperation from personnel are important.
4. Technical assistance and guidance on the ABC methodology may be necessary initially.
5. To derive long-term benefits, cost data trending is essential. Data trending requires continued efforts of costing, which in turn requires management commitment.

2.4.4 ABC IN GOVERNMENTS

State and local governments have a common goal: to provide services to the public with acceptable quality and at the lowest possible cost. Many governments, however, are not perceived to be particularly efficient in realizing this goal. Most governments do not have a clear idea of all the costs associated with their own in-house operations. They typically underestimate the true costs of in-house operations by as much as 30 percent, typically because they omit many indirect costs when determining the total costs of performing any function.[9]

Traditionally, state and local governments have practiced cost control by simply aggregating costs for the units within a governmental body and comparing the total costs of these units from period to period. Overhead costs that applied to multiple units were normally allocated to the units based on arbitrary measurements that are common to all units, such as square feet occupied or DL costs. Volume-based measurements are used predominantly in traditional systems to determine overhead rates and to assign overhead costs to their activities.[10]

The Texas Department of Agriculture (TDA) operates six livestock facilities for inspecting animals prior to exporting them to Mexico. The inspection is necessary in order to be in compliance with Mexico's health regulations. The TDA charges fees for the inspections. When trucks carrying livestock arrive at the TDA export pens, the pen manager checks the driver's document before authorizing workers to unload the truck. The unloaded animals are placed in pens to rest. They are then moved to an inspection area and then inspected by a veterinarian. Animals that do not pass inspection are reloaded, fees are collected, and export to Mexico is denied. Those animals that do pass inspection are immediately reloaded or returned to the initial pens to wait for a truck. Once the truck has been reloaded and cleared for export, required fees are collected, the document is returned to the driver, and the truck is sealed for departure to Mexico. Four steps are taken to apply ABC to this case:

1. Identification of costs and resources
2. Identification of the direct and indirect costs of activities
3. Assignment of costs to activities
4. Calculation of unit costs

There are altogether nine distinct activities. Costs are grouped into four primary categories and one indirect cost category. Five cost pools are created. The total costs of each pool were expressed on a per-unit basis according to the appropriate driver for the particular activity. For details of its quantitative analyses, see Briner et al.[11]

2.4.5 *ABC IN SOFTWARE DEVELOPMENT*

Traditional software development follows a "waterfall" approach by performing needs analysis, coding software, testing, documentation, training, and implementation. Component-based software development, on the other hand, aspires to create reusable codes on a large scale. The overhead costs for developing component-based codes are expected to be significant due to the maintenance of reuse infrastructure and the development, location, evaluation, and adaptation of components. Table 2.11 lists some of the activities that are to be cost estimated and allocated to the proper cost objects in such an environment using the ABC method.[12]

In Table 2.12 the corresponding cost drivers are identified to implement the ABC costing system. Here, the basic steps of applying ABC are quite similar to those in other service sectors.

In summary, it is worth noting again that the ABC technique is capable of redistributing the overhead portion of the total costs to various cost objects (e.g., products and customers). Its application does not reduce cost or increase profits. However, the outcome of an ABC analysis produces insights regarding relative profitability that could be useful as a basis for new corporate strategy (e.g., pricing, emphasis placed on different customer groups, etc.). For companies to improve profitability, such new corporate strategies need to be effectively implemented.

2.5 TARGET COSTING

Target costing is a technique for setting the upper bound for the unit cost of a new product/service, above which the new product/service may no longer be profitable for the company to pursue.

If the products or services offered by companies are new and without competition, companies can easily price them on the basis of *cost plus*, which means defining product prices by adding an acceptable gross margin on top of their product costs. On the other hand, for products and services offered to a competitive marketplace, prices may need to be constantly revised due to external market forces (i.e., competition, economy, technologies, industrial status, etc.). In such cases, companies are able to maintain their gross margins only if they are in a position to adjust their

Table 2.11. Software development activities to be cost estimated using ABC

Activity center	Activity	Description
A Reuse infrastructure creation	1. Develop reuse infrastructure	Develop reuse policies, tools, processes, and measures
B Reuse infrastructure maintenance and reuse marketing	2. Maintain reuse infrastructure	Maintain reuse policies, tools, processes, and measures
	3. Communicate existence of components	Advertise the reuse components
C Reuse administration	4. Administer reuse measurement accounting and incentives	Measure ongoing levels of reuse on projects and allocate costs of reuse. Reward individuals or teams, or both, for reuse achieved.
D Reuse production	5. Analyze reuse components	Investigate opportunities for acquiring for developing components
	6. Develop or acquire reuse components	Develop components to be reusable, acquire or generalize, or both, components previously developed
	7. Certify components	Certify components for reusability
	8. Document, classify, and store components	Document, classify, and store offered components
E Reuse consumption	9. Search for components	Search libraries for components (or assist the search process)
	10. Retrieve, understand, and evaluate components	Understand and evaluate components found in libraries
	11. Adapt and integrate components	Modify (if necessary) and integrate components

(*Continued*)

Table 2.11. (*Continued*)

Activity center	Activity	Description
F Reuse maintenance	12. Maintain reusable components	Correct and extend components in the library
	13. Update reusable components	Correct and extend components in systems.

Source: Fichman and Kemerer (2002).

costs downwards according to the externally imposed price reductions. Target costing represents a useful management tool to assist companies in such circumstances.[13] The principles of target costing are as follows:

1. *Price-led costing*: Market prices are used to determine allowable—or target—costs. Target costs are calculated using a formula similar to the following: market price—required profit margin = target cost.
2. *Focus on customers*: Customer requirements for quality, cost, and time are simultaneously incorporated in product and process decisions and these requirements guide cost analysis. The value (to the customers) of any features and functionality built into the product must be greater than the cost of providing those features and functionality.
3. *Focus on design*: Cost control is emphasized at the product and process design stage. Therefore, engineering changes must occur before production designs, resulting in lower costs and reduced "time-to-market" for new products.
4. *Cross-functional involvement*: Cross-functional product and process teams are responsible for the entire product from initial concept through final production. They would hold brainstorming sessions to generate ideas to cut costs.
5. *Value-chain involvement*: All members of the value chain (e.g., suppliers, distributors, service providers, and customers) are included in the target costing process, as well as any cost-saving opportunities offered by suppliers or based on customer inputs.
6. *A life-cycle orientation*: Total life-cycle costs are minimized for both the producer and the customer. Life-cycle costs include purchase price, operating costs, maintenance costs, and distribution (disposal) costs.

Companies use target costing to establish concrete and highly visible cost targets for their new products/services. The gaps between the target cost and cost projections for the new product are to be minimized using innovations and basic value engineering tools.

Table 2.12. Cost drivers in developing reusable software

Activity	Casual cost driver	Performance measure
1. Develop reuse infrastructure	Reuse policy and expected volume	Breadth and quality of infrastructure
2. Maintain reuse infrastructure	Size/complexity of infrastructure	Breadth and quality of infrastructure
3. Communicate existence of components	Size of reuse libraries. Size of developer community	Number of communications
4. Administer reuse measurement accounting and incentives	Reuse policy. Volume of reuse	Number of measurements
5. Analyze reuse components	Reuse policy	Number of components evaluated
6. Develop or acquire reuse components	Reuse opportunity	Number and quality of reusable components acquired
7. Certify components	Offered components	Number of components certified
8. Document, classify, and store components	Certified components	Number of components stored
9. Search for components	Project policy. Reuse potential of the domain. Size/quality of reuse libraries	Number of searches Number of promising components found
10. Retrieve, understand, and evaluate components	Promising components found	Number of components evaluated and number of components accepted
11. Adapt and integrate components	Retrieved components selected for use	Number of attempted integrations; number of successful integrations
12. Maintain reusable components	Volume of reuse	Number of components maintained
13. Update reusable components	Volume of reuse	Number of components updated

2.6 RISK ANALYSIS AND COST ESTIMATION UNDER UNCERTAINTY

When estimating product costs, some costs are well defined and firm, while others are not. Similarly, some engineering projects built from past experience are more or less risk free, while others are not. Risks are defined as a measure of the potential variability of an outcome (e.g., cost or schedule) from its expected value. Risks must be properly accounted for in projects.[14] Various application examples are included in Akira.[15]

2.6.1 MATHEMATICAL REPRESENTATION OF RISKS

Risks can be graphically represented by a probability distribution function. Three cases are examined next in Figure 2.2.

Case 1 refers to an investment in U.S. Treasury bills. The yield of 10-year U.S. Treasury bills has varied in the range of 8 percent (1990) to below 2 percent (2013). However, once an investor purchases the Treasury bills from the U.S. Government, the yield is locked in until maturity. Such an investment is guaranteed by the assets of the U.S. Government. This return is graphically represented by a vertical line at a fixed return rate (6 percent assumed in this example) with 100 percent probability.

Case 2 is an investment in a blue-chip corporate stock with a most likely return of 8 percent. Due to market conditions being usually unpredictable, the return of such an investment has some measure of risk, as represented by the bell-shaped curve centered on 8 percent in Figure 2.2. The return may vary from 4 percent (minimum) to 12 percent (maximum). Risks are measured by the standard deviation of this probability distribution curve (e.g., σ_2).

The bell-shaped curve is mathematically represented by the normal probability density function:

$$F(x) = 1/(2\pi)\text{\^{}}0.5\ (1/\sigma)\ \text{EXP}\ [-0.5\ ((x-\mu)/\sigma)\text{\^{}}2] \tag{2.1}$$

where σ = standard deviation;
μ = mean; π = 3.14159;
^ = superscript. For example: *A^b* means *A* (sup) *b*.

The area underneath the curve is normalized to be 1.

Case 3 is the return of an investment in real estate, centered, for example, around 15 percent. Because this investment requires tax

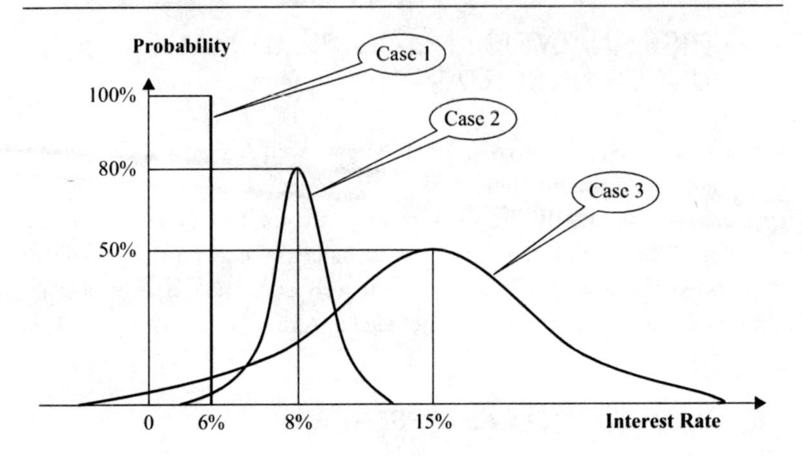

Figure 2.2. Representation of risks.

payment, maintenance costs, and other expenditures, its minimum return may be negative. Its upside potential may be very large, however, if commercial developments and property-zoning results become favorable. This case is represented by a bell-shaped probability curve having a large standard deviation (σ_3 being larger than σ_2) and with its most likely return centered around 15 percent. Furthermore, the probability of achieving its most likely return of 15 percent is now only 50 percent (see Figure 2.2).

Risky events may be represented mathematically by the normal probability density function, which is defined by two parameters, standard deviation and mean. Besides the normal, several other probability density functions (such as triangular, Poisson, and beta) may also be used to represent risky costs (see Section 2.6.3).

2.6.2 PROJECT COST ESTIMATION BY SIMULATION

Recent literature outlines the advancements of PC-based techniques that estimate project costs under uncertainty.[16] The key steps of applying these PC-based techniques are as follows:

1. Construct a *cost model* for the projects at hand with a spreadsheet program (e.g., Excel®).[17] The spreadsheet program takes care of the required computation steps of the cost model, such as addition, subtraction, multiplication, and division. The numerical values entered in the spreadsheet cells are typically deterministic, each having a

well-defined and fixed magnitude. The cost model encompasses all cost components and computes the total project cost.

2. Make a *three-point estimate* for each of the component costs, composed of the minimum, the most likely, and the maximum values. This is to account for the perceived cost uncertainty. Past experience may serve as a guide in the selection of these values.

   Select a probability distribution function (e.g., triangular, normal, beta, or other distribution functions) to represent the three-point estimate of the component cost. Repeat this step for all other cost components of the project.

3. Activate *risk analysis software* to replace the deterministic values contained in the spreadsheet cells by the probability distribution functions chosen to represent its corresponding three-point estimates. Currently, commercial PC-based software products are becoming readily available. A specific application example is "CrystalBall®" for Excel.[18]

   The activated risk analysis software automatically converts all input probability density functions to their corresponding cumulative distribution functions. The technical fundamentals related to this conversion are illustrated in Appendix 2.10.5.

4. Conduct *Monte Carlo Simulations* to compute the total project cost. Upon activation of risk analysis software, a random number is first generated between 0 and 1. This random number is a trial probability value (e.g., P1 in Figure 2.3). Using this random number, the specific cost value is read from the cumulative distribution of the cost component C1, which represents a random input variable (e.g., C11 in Figure 2.3). A second random number is generated (P2), which is then used to define the cost components C2 (e.g., C21). A third random number is generated to define the cost C31 of a third cost component. This process is continued until the costs of all cost components are defined. The total project cost (e.g., TPC1) is then calculated with the spreadsheet program that contains the cost model. This is one outcome of the random output variable TPC.

   The sampling process is repeated thousands of times to create a distribution of the total project costs (e.g., TPC1, TPC2, . . .). These output results are then statistically grouped into bins (with zero to maximum value) to define a cumulative distribution. The resulting cumulative distribution for the total project cost TPC may then be converted back to its corresponding probability density function.

   The total project cost so generated has a set of minimum, most likely, and maximum values.

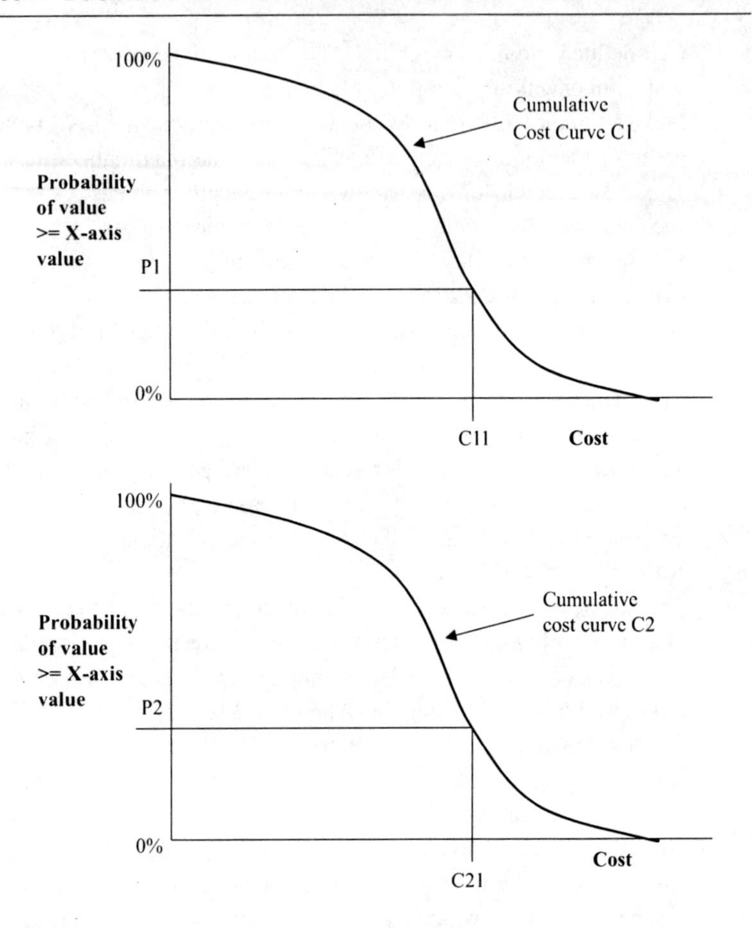

Figure 2.3. Cumulative distribution functions.

5. Interpret the *total project cost* represented in a cumulative distribution to arrive at the following typical results:
   - There is an 80 percent probability that the total project cost will not exceed \$*D*.
   - The minimum, most likely, and maximum total project costs are \$*A*, \$*B*, and \$*C*, *respectively*.
   - The standard deviation of the total project cost is σ, which represents the overall measure of the project risk.

Results of this type are garnering excellent reception in industry as they inform better decision making. It is particularly true in situations where multiple projects are being evaluated for investment purposes.

There is an additional benefit which is realizable by using the just-described cost estimation method by simulations. Because of the "risk pooling" effect due to risk sharing among all input cost components, the total project cost is expected to have a lower overall risk than the risk levels of its individual components. Various studies (e.g., Canada et al.[19]) have confirmed that the total project cost computed by simulations requires a smaller contingency cost for a given risk level than that computed by the traditional method of using deterministic values. Other important applications involving risk analyses include (a) project schedule and (b) portfolio optimization.

The value of risk analyses is typically to make explicit the uncertainties of input variables, to promote more reasoned estimating procedures, to allow more comprehensive analyses—or the simultaneous variation of all input variables involved—and to measure the variability of output variables. There is a plethora of evidence that a decision maker will make better decisions with a fuller understanding of the risk-based implications of all input variables involved.

The use of risk analysis in the business and engineering environments is expected to become increasingly widespread in the years to come. Engineering managers are advised to become familiar with such advanced tools for risk analyses.

2.6.3 EXAMPLES OF INPUT DISTRIBUTION FUNCTIONS

In engineering cost estimation, several distribution functions are often used as inputs. Figure 2.4 shows the triangular probability density function. This is the easiest function to apply, as the three-point estimates may be directly incorporated into this representation scheme.

Figure 2.5 illustrates the normal probability function. Figure 2.6 displays the beta probability density function. Figure 2.7 depicts the Poisson probability function.

Any of these probability density distribution functions may be used to represent the input values for given cost-estimation project.

2.6.4 APPLICATION—COST ESTIMATION OF A RISKY CAPITAL PROJECT

As an example, the cost estimation for a turnkey capital project is illustrated in Table 2.13. Project managers define the base (e.g., the most likely) estimates, as well as the lower and upper bounds for each cost item in the estimate. Doing so will force them to externalize the reasons for any variance and require them to think hard about the contingency plans for each.

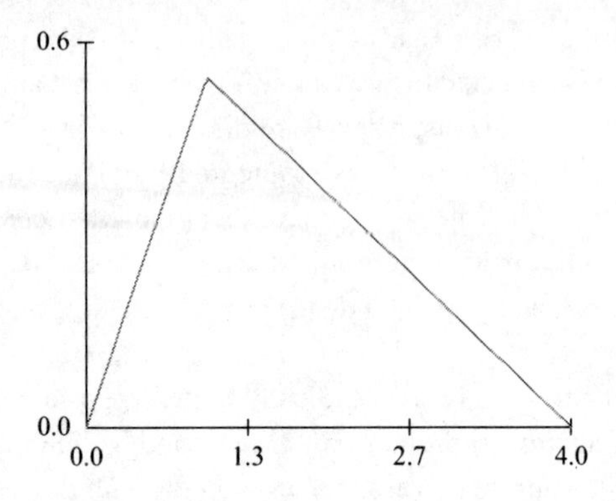

Figure 2.4. Triangular probability density function.

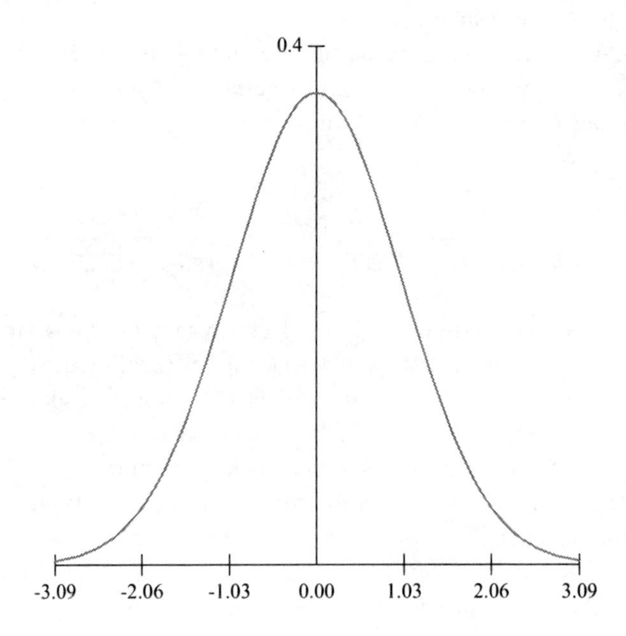

Figure 2.5. Normal probability density function.

The output total cost is represented by the probability density and cumulative distribution functions (see Figures 2.8 and 2.9). From the output cumulative distribution, the following results are readily derived:

1. The most likely total project cost is $5,136,000, which, of course, only echoes the input data.

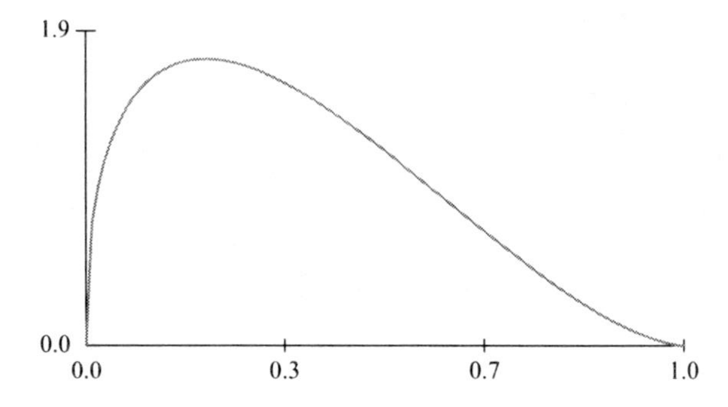

Figure 2.6. Beta distribution function.

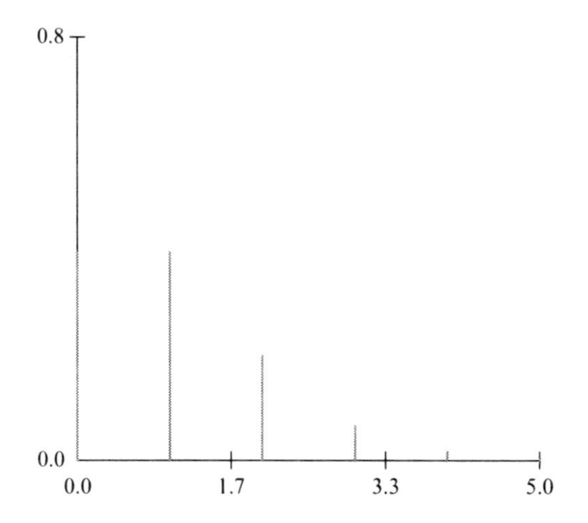

Figure 2.7. Poisson distribution functions.

2. There is an 80 percent probability that the project cost will exceed $5,100,000.
3. There is a 20 percent probability that the project cost will exceed $5,170,000.
4. The maximum project cost is $5,250,000.
5. The minimum project cost is $4,989,710
6. The standard deviation cost (σ) is $62,172.

Note that the information offered by items 2, 3, 4, 5 and 6 are new. In a traditional, deterministic project-cost estimate, the cost figures for items 1 would be the same, and there would be no information of the type offered by the remainder. When choosing among various projects that may have similar outcomes in the most likely project costs, information offered

Table 2.13. Cost model of a capital project (thousands of dollars)

#	Cost Category	Base ($K)	Min	Max	Minimum ($)	Most likely ($)	Maximum ($)
1XXX	Cold box	$748.00	–5%	5%	$710.60	$748.00	$785.40
2XXX	Rotating equipment	$742.00	–3%	3%	$719.74	$742.00	$764.26
3XXX	Process equipment	$658.00	–2%	5%	$644.84	$658.00	$690.90
4XXX	Electrical equipment	$194.00	–5%	3%	$184.30	$194.00	$199.82
5XXX	Instrumentation	$295.00	–2%	10%	$289.10	$295.00	$324.50
6XXX	Piping mat/specials	$121.00	–2%	10%	$118.58	$121.00	$133.10
711X	Civil construction	$284.00	–2%	5%	$278.32	$284.00	$298.20
712X	Mechanical construction	$390.00	–1%	20%	$386.10	$390.00	$468.00
713X	Electrical construction	$85.00	–2%	10%	$83.30	$85.00	$93.50
71?X	Other contracts	$83.00	–5%	10%	$78.85	$83.00	$91.30
716X	Purchased enclosures	$48.00	–5%	12%	$45.60	$48.00	$53.76

717X	Fabrication	$179.00	–5%	4%	$170.05	$179.00	$186.16
7890	Freight	$80.00	–5%	15%	$76.00	$80.00	$92.00
84X0	Field support	$188.00	–5%	7%	$178.60	$188.00	$201.16
85XX	Start-up	$60.00	–10%	30%	$54.00	$60.00	$78.00
81X0	Product line design	$516.00	–1%	15%	$510.84	$516.00	$593.40
8150	Project execution	$333.00	–5%	20%	$316.35	$333.00	$399.60
	Total neat	$5,004.00					
	Contingency	$131.90					
	Grand total	**$5,135.90**					

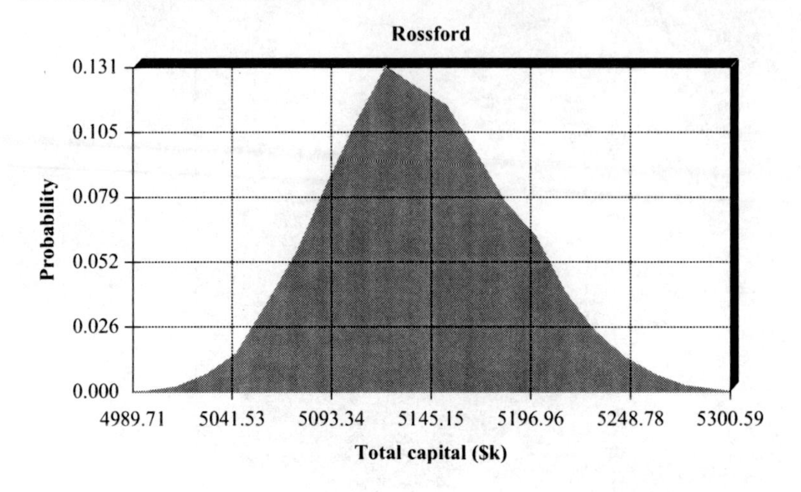

Figure 2.8. Total capital cost presented in a probability density function.

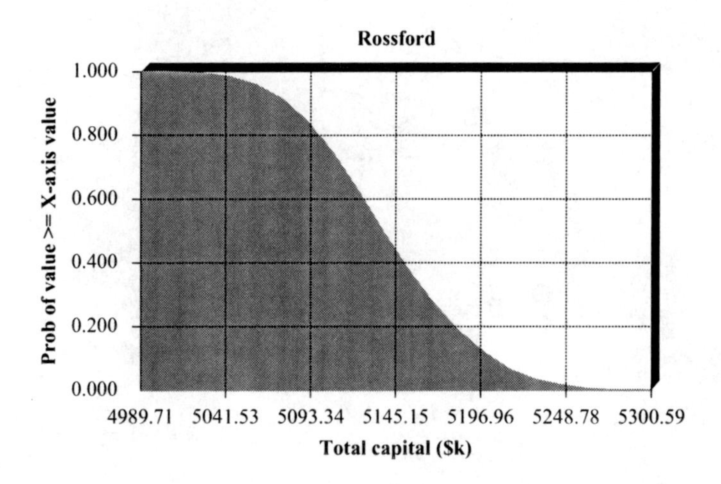

Figure 2.9. Total capital cost presented in a cumulative distribution function.

by items 2 and 3, 4, 5 and 6 is especially critical for differentiating projects by their inherent risks.

For construction managers, the estimation of contingency is of critical importance. Instead of assigning specific contingencies as a percentage to each cost category item, the simulation technique calculates the contingency of the construction project. Probabilistic models have been used in the past to define construction project contingency.[20]

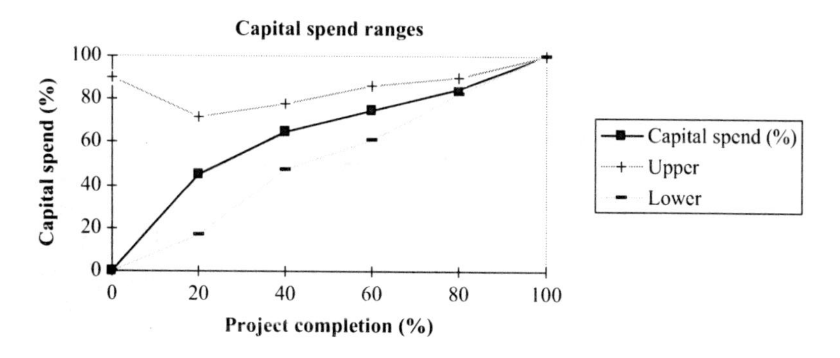

Figure 2.10. Project control and tracking.

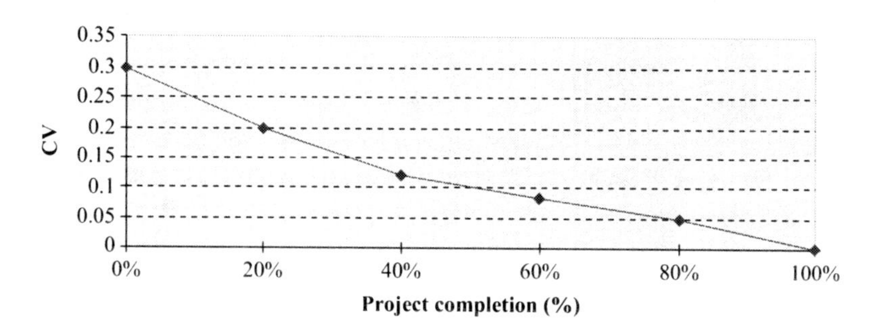

Figure 2.11. Coefficient of variation.

In addition, project control and tracking of coefficient of variation can be readily conducted (see Figures 2.10 and 2.11). The coefficient of variation is defined as CV = 100 (σ/μ), wherein σ is the standard deviation and μ is the mean of the total project cost distribution function.

2.6.5 OTHER TECHNIQUES TO ACCOUNT FOR RISKS

Several other techniques are also routinely applied in industry to assess and manage the risks associated with projects of various kinds.[21]

1. **Sensitivity analysis.** Because of possible variation of specific input parameters, "what-if" analyses are typically performed to assess the sensitivity of the project cost and time to completion.
2. **Contingency cost estimation.** The cost of a risky project may be estimated by adding an empirical (based on best practices)

contingency cost to each task (typically 5–7 percent of the task cost) to cover the risk involved.

3. **Decision trees.** Decision trees are used to evaluate sequential decisions and decide on alternatives on the basis of the computed expected values of probabilistic outcomes (see Example 2.2).
4. **Diversification.** With this risk-management technique, several risky projects can be engaged at the same time to spread the risks by means of diversification—combining high-risk, high-return projects with low-risk, low-return projects—to achieve a reasonable overall return on investment.
5. **Fuzzy logic systems.** These systems of reasoning are based on fuzzy sets.[22] A fuzzy set defines the range of values for a given concept as well as the degree of membership. A membership of 1 indicates full membership, whereas 0 defines exclusion. The change of membership from 0 to 1 is gradual. For example, a fuzzy expert system employs rules such as the following: If the temperature *T* is high (Ai) and the difference in temperature is small (Bi), then close the valve *V* slightly (Si). Here Ai, Bi, and Si are fuzzy sets. Fuzzy logic systems have been applied to assess project risks.[23]

Example 2.2

The company needs to decide either to develop a new product with an investment of $400,000 or to upgrade an existing product by spending $200,000, as illustrated by the decision point 1 in Figure 2.12.

A. If the new product strategy is pursued, then there is a 60-percent probability that the product will be in high demand, in which case the company will make $200,000 next year. Concurrently, there is a 40-percent chance for low demand, which will result in a loss of $100,00 for the company next year (see Node A in the decision tree diagram in Figure 2.12).

   If the new product enjoys a high demand, then there is an 80-percent chance that the product will make $1,000,000 and a 20-percent chance it will make only $50,000 in the year after the next (see Node a in Figure 2.12). If the new product meets a low demand, then there is a 30-percent chance of a revenue of $500,000 and a 70-percent chance of suffering a loss of $500,000 in the year after next (see Node b in Figure 2.12).

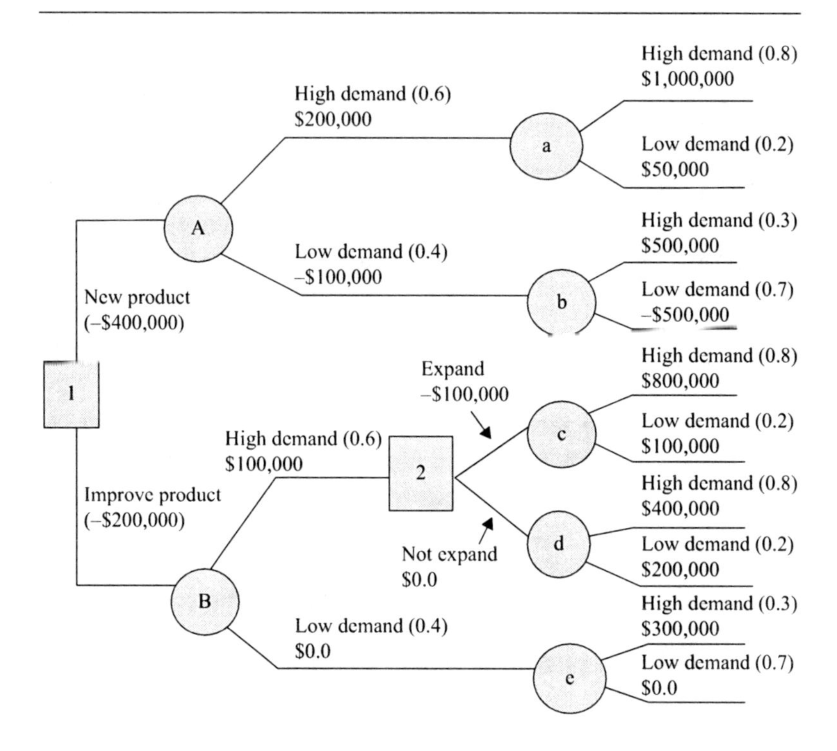

Figure 2.12. Decision tree analysis.

B. If the company follows the strategy of improving the existing product, then there is a 60-percent chance for high demand, leading to a revenue of $100,000; and a 40-percent chance of low demand, to yield zero revenue in the first year (see Node B in Figure 2.12.)

If the demand is high at the end of the first year, the company needs to make a second decision (Decision Point 2) whether or not to expand the product line. The expansion option will require an investment of $100,000, whereas the option of no expansion costs nothing. If the company expands the improved product line, then there is an 80-percent chance that it will reap a revenue of $800,000 and a 20-percent chance of $100,000 revenue in the year after the next (see Node c in Figure 2.11). If the company elects not to expand at the end of the first year, then there is an 80-percent chance that it will realize a revenue of $400,000 and a 20-percent chance of a $200,000 revenue (see Node d in Figure 2.12).

Should the improved product of the company see a low demand (at a probability of 40 percent) and get zero revenue in the first year, then there is a 30-percent chance that the product can generate a revenue of $300,000 and a 70-percent chance of zero revenue (see Node e in Figure 2.12).

The interest rate is 10 percent. This problem is fully diagrammed in Figure 2.12. Determine which decisions at Decision Points 1 (product strategy) and 2 (expansion strategy) should the company make.

Answer 2.2

This problem can be solved by using the decision tree method, which works from right to left, from future (the year after next) to present.

Let us define: ER = expected return and PV = present value.

A. Decision Point 2.

At Node c, the expansion option has a total expected return of

$$ER(c) = 0.8\ (\$800{,}000) + 0.2\ (\$100{,}000) = \$660{,}000$$

At Node d, the no-expansion option has a total expected return of

$$ER(d) = 0.8\ (\$400{,}000) + 0.2\ (\$200{,}000) = \$360{,}000$$

The present value of expansion is

$$PV(\text{Expansion}) = -\$100{,}000 + ER(c)/1.1 = \$500{,}000$$

And the present value of no expansion is

$$PV(\text{No expansion}) = 0 + ER(d)/1.1 = \$327{,}272$$

Based on these present values, the decision should favor expansion.

B. Decision Point 1.

At Node a, the total expected return is

$$ER(a) = 0.8\ (\$1{,}000{,}000) + 0.2\ (\$50{,}000) = \$810{,}000$$

At Node b, the corresponding total expected return is

$$ER(b) = 0.3\ (\$500{,}000) + 0.7\ (-\$500{,}000) = -\$200{,}000$$

Thus, the present value for the high demand case is

$$\text{PV(High Demand)} = \$200{,}000 + \text{ER (a)}/(1.1) = \$936{,}363$$

And the present value for the low demand case is

$$\text{PV(Low Demand)} = -\$100{,}000 + \text{ER (b)}/1.1 = -\$281{,}818$$

Thus, the present value for the new product strategy is:

$$\text{PV(New Product)} = -\$400{,}000 + [0.6\ \text{PV(High Demand)} + 0.4\ \text{PV(Low Demand)}]/1.1 = \$8{,}264$$

On the other hand, we have the following for the product improvement strategy:

$$\text{PV(High Demand)} = \$100{,}000 + \text{PV(Expansion)} = \$600{,}000$$

Note that the no-expansion option is abandoned.
The present value of low demand is

$$\text{PV(Low Demand)} = 0 + \text{ER(e)}/1.1 = 0 + [0.3(\$300{,}000) + 0.7(0)]/1.1 = \$81{,}818$$

Thus, the present value for the product improvement strategy is

$$\text{PV(Product Improvement)} = -\$200{,}000 + [0.6(\$600{,}000) + 0.4(\$81{,}818)]/1.1 = \$157{,}024$$

Since the present value for improving the product is larger than the present value for new product, the choice should be in favor of product improvement.

Even if the company decides to forgo expansion at Decision Point 2, the present value for product improvement is

$$\text{PV(Product Improvement with no expansion)} = -200{,}000 + [0.6(\$427{,}272) + 0.4\ (\$81{,}818)]/1.1 = \$62{,}809$$

which is still larger than that for new product development. The value of expansion is thus:

$$\text{Value (Expansion)} = \$157{,}024 - 62{,}809 = \$94{,}215.$$

2.7 REDUCTION OF OVERHEAD COSTS

Some of the overhead costs are related to finance, human resources, IT, and legal functions. Overhead costs are known to grow directly with gross margins and they often grow faster than revenues. Several best practices are noted in the literature to reduce these overhead costs,[24] such as:

A. Aligning the overhead capacities with the company's strategic needs:
    1. Streamline the planning process for marketing
    2. Use enterprise resource planning tools
    3. Outsource IT support
    4. Automate/centralize cash incentive preparation

B. Reducing the needs for these overhead functions:
    5. Change reporting from monthly to quarterly
    6. Change payroll from weekly to biweekly
    7. Eliminate nonstrategic technology initiatives
    8. Replace quarterly internal strategy meeting with monthly conference calls
    9. Eliminate statistical analysis group
    10. Reduce the frequency of long-term trend analysis

C. Optimizing the delivery of these overhead activities:
    11. Standardize reporting format and eliminate redundant ones
    12. Enforce purchasing standards
    13. Set travel and expenses standards
    14. Use self-service for human resources (HR) benefits
    15. Use temporary contract personnel
    16. Consolidate advertisement agency work and eliminate high-cost outliers

Tom Peters said: "Almost all quality improvement comes via simplification of design, manufacturing, layout, processes and procedures." Simplification represents a useful strategy to reduce costs. Since overhead costs may represent a large percentage of the overall cost of doing business, it is important for engineering managers and professionals to constantly look for ways to reduce these costs.

2.8 MISCELLANEOUS TOPICS

This section discusses several miscellaneous topics, including simple cost-based decision models, and project evaluation criteria.

2.8.1 *SIMPLE COST-BASED DECISION MODELS*

Engineering managers need to make regular choices among alternatives. In some cases, such choices may be made based on costs, as illustrated by the following two examples.

A. ***Comparison of alternatives.*** When faced with the option of purchasing one of several sets of capital equipment with similar functional characteristics, engineering managers can use the following annual cost (*AC*) formula to identify which has the lowest total *AC*.

The *AC* for a long-lived asset is defined as the sum of its depreciation charge, interest charge for the capital tied down by the purchase, and its annual operational expenses; that is,

$$AC = \frac{(P - L)\times i}{[(1+i)\wedge N - 1]} + P \times i + AE \qquad (2.2)$$

where P = initial investment ($)
L = salvage value ($)
N = useful life of a long-lived asset (year)
i = interest rate (%)
AE = annual expenses ($)—taxes, supplies, insurance repairs, utilities, and so on.
AC = annual cost ($)

The capital equipment with the lowest *AC* is to be preferred.

Example 2.3

Your company has averaged 15-percent growth per year for the past 7 years, and now you need additional warehouse space for purchased material as well as FG. Two types of constructions have been under consideration: conventional and air-supported fabric. (The data are available in Table 2.14). Which is the better economical choice?

Answer 2.3

Conventional:

$AC1$ = (200,000 – 40,000) × 0.08/[1.08^40 – 1] + 200,000 × 0.08 + (1500 + 700 + 1.5 × 200,000/100) = $21,818

Table 2.14. Warehouse options

Type	**Conventional**	**Air-supported**
First cost ($)	200,000	35,000
Life (years)	40	8
Annual maintenance ($)	1,500	5,000
Power and fuel ($)	700	5,500
Annual taxes ($)	1.5/100	1.5/100
Salvage value ($)	40,000	3,000
Interest rate	0.08	0.08

For air-supported:

$$AC2 = (35{,}000 - 3{,}000) \times 0.08/[1.08\text{^}8 - 1] + 35{,}000 \times 0.08 + (5000 + 5500 + 1.5 \times 35{,}000/100) = \$16{,}833$$

Choice: air-supported system.

Example 2.4

A semiautomatic machine is quoted at $15,000, while an advanced machine is quoted at $25,000. The salvage value of these machines is assumed to be zero. A four-man party can produce 500 parts a day with the semiautomatic machine, by using a machinist at $200 per day, a maintenance worker at $150 per day, a parts laborer at $100 per day, and a warehouse clerk at $80 per day.

A six-man party can produce 770 parts a day with the advanced machine, using a machinist at $200 per day, an assistant machinist at $180 per day, a maintenance worker at $150 per day, two parts laborers at $100 per day each, and a warehouse clerk at $80 per day.

Material cost is $10 per part. The FO is 50 percent of the DL cost, only when parts are being produced. The maintenance expense for the semiautomatic machine is $250 per year, and for the advanced machine it is $500 per day. The estimated life of the semiautomatic machine is 20 years, and 15 years for the advanced machine. The interest rate is 8 percent per year.

How many parts must be made per year to justify the deployment of the advanced machine?

Table 2.15. Comparison of two machines

Items of interest	Semiautomatic	Advanced
First cost ($)	15,000	25,000
Daily production	500	850
Daily wage ($)	530	810
Annual maintenance ($)	250	500
Life (years)	20	15
Material cost ($/part)	10	10
Factory overhead as percent	50%	50%
Interest	0.08	0.08

Answer 2.4

Define N = number of parts produced per year.

y = number of working days to produce parts. For the semiautomatic machine, $y = N/500$; for the advanced machine, $y = N/770$ (see Table 2.15).

The AC of operating the semiautomatic machine is given by:

$$AC\,(SA) = (15{,}000 \times 0.08)/[(1.08)\text{^}20 - 1] + 15{,}000 \times 0.08 + 250 + 10 * N + 530\,(N/500) \times 1.5$$
$$= 1777.78 + 11.59N$$

The AC of operating the advanced machine is given by:

$$AC\,(A) = (25{,}000 \times 0.08)/[(1.08)\text{^}15 - 1] + 25{,}000 \times 0.08 + 500 + 10 * \mathrm{N} + 810\,(N/770) \times 1.5$$
$$= 3420.74 + 11.57792N$$

Setting $AC\,(SA) = AC\,(A)$, we have:

$$1775.78 + 11.59N = 3420.74 + 11.57792N$$
$$N = 136{,}172$$

The advanced machine is justifiable, if the production exceeds 136,172 parts per year.

B. ***Replacement evaluation.*** Engineering managers are sometimes faced with the decision of whether to replace an existing facility with a brand-new one. Again, this replacement decision may be made by identifying the option with the lowest *AC*.

In this analysis, the existing facility is treated as if it is new, in that its residual equipment life and its residual book value (initial capital investment minus accumulated depreciation) are equivalent, respectively, to the useful product life and the capital investment cost of new equipment. Thus,

$$AC_0 = \frac{i \times [BV(t) - L_0]}{(1+i)^\wedge(N_0 - t) - 1} + BV(t) \times i + AE_0 \tag{2.3}$$

$$AC = \frac{i \times [P - L]}{(1+i)^\wedge N - 1} + P \times i + AE \tag{2.4}$$

where

P_o = original investment cost ($)
N_o = original estimate of useful life (year)
AE_o = annual expenses of using the original equipment ($)
L_o = salvage value at the end of the equipment's original useful life ($)
AC_o = Annual cost for using the existing equipment ($)
t = Present age of equipment (year)
$N_o - t$ = remainder life (year)
$BV(t)$ = book value of existing equipment at the end of the t(th) year ($) = P_o - accumulated depreciation at the end of the t(th) year
P = initial investment of the replacement equipment ($)
AC = annual cost of using the replacement equipment ($)
L = Salvage value of the replacement equipment at the end of its useful life (year)
N = useful life of the replacement equipment (year)
AE = annual expenses for using the replacement equipment ($)

If $AC_o > AC$, use the replacement equipment to save costs.

Example 2.5

A compressor air-supply station was built 18 years ago at the main shaft entrance to a coal mine at a cost of $2.6 million. The station was equipped with steam-driven air compressors that have an annual operating expense of \$360,000. The salvage value at the estimated 25-year life of the station is \$130,000. It can be sold now for \$800,000.

A proposal has been made to replace the station with electrically driven compressors that would be installed underground near the working face of the mine for a cost of \$2.8 million. The new compressor station would have a life of 30 years and a salvage value of 10 percent of its initial cost. Its annual operating cost would be two-thirds of that of the steam-driven station. Annual taxes and insurance are 2.5 percent of the first cost of either station. The interest rate is 8 percent. Is there a financial justification to replace the steam station?

Answer 2.5

Table 2.16 summarizes the data of these two compressors.

Assume that we, use the exact method for calculating depreciation (sinking fund):

$$\text{Steam: AC} = (800,000 - 130,000) \times \frac{0.08}{1.08^7 - 1} + 800,000 \times 0.08 + 425,000 = 564,088.50$$

$$\text{Electric: AC} = \frac{(2,800,000 - 280,000) \times 0.08}{1.08^{30} - 1} + 2,800,000 \times 0.08 + 310,000 = 556,245.13$$

Using the exact method, the conclusion reached is to replace the old steam unit with electrically driven compressors.

Table 2.16. Comparison of two compressor drives

Available data	Steam	Electric
Original investment (\$)	2,600,000	2,800,000
Life (years)	25	30
Present age (years)	18	0
Remaining life (years)	7	30
Present salvage value (\$)	800,000	
Final salvage value (\$)	130,000	280,000
Annual expense (\$)	360,000 + 65,000	240,000 + 70,000

2.8.2 PROJECT EVALUATION CRITERIA

Engineering managers are often required to make choices among capital projects that may deliver benefits and at the same time consume resources on an annual basis over a number of periods. Several standard methods are used in industry to evaluate such projects. These include *net present value (NPV), internal rate of return (IRR), payback, and profitability index (PI).*

A. Net present value

$$NPV = -P + \sum \frac{NCIF(m)}{(1+i)^\wedge m} + \frac{CR}{(1+i)^\wedge n} \tag{2.5}$$

$m = 1$ to n

where

NPV = net present value (dollars)

P = present investment made to initiate a project activity (dollars).

NCIF (m) = net cash in-flow (dollars) in the period m, which represents revenues earned minus costs incurred, (R(m) – C(m)) in dollars

i = cost of capital (interest rate)

n = number of interest period (year)

CR = capital recovery (dollars), which is the amount regained at the end of the project through resale or other methods of dispositions

Σ = the summation sign, for m = 1 to n.

Note that the first term on the right-hand side of Equation 2.5 is the capital outlay for the project, or an outflow of value (cash). The second term on the right-hand side is the sum of discounted net cash in-flow earned over the years. The third term on the right-hand side is the discounted capital recovery of the project.

One major weakness of the NPV equation is that all benefits derived from a project must be expressed in dollar equivalents—within NCIF(m)—in order to be included. Nonmonetary benefits, such as enhanced corporate image, expanded market share, and others, cannot be represented.

For the special case of NCIF(m) = CF = constant

$$NPV = -P + CF\frac{(1+i)^\wedge n - 1}{i \times (1+i)^\wedge n} + \frac{CR}{(1+i)^\wedge n} \tag{2.6}$$

Projects with the largest NPV values are preferable, as NPV represents the net total value added (before tax) to the firm by having completed the project at hand. Note that NPV may be determined only if the project's net cash in-flow NCIF(m) is known.

Example 2.6

The company is evaluating two specific proposals to market a new product. The current interest rate is 10 percent. Proposal A calls for setting up an in-house manufacturing shop to make the product, requiring an investment of $500,000. The expected profits for the first to fifth years are $150,000, $200,000, $250,000, $150,000, and $100,000, respectively. Proposal B suggests that the manufacturing operation be outsourced by contracting an outside shop, requiring a front-end payment of $300,000. The expected profits for the first to fifth years are $50,000, $150,000, $200,000, $300,000, and $200,000, respectively. The expected profits would be lower in earlier years due to third-party markup. There is no salvage value which can be recovered at the end of either proposal. Which proposal should the company accept?

Answer 2.6

The NPV equation should be used to evaluate these two proposals:

$$\text{NPV (A)} = -500{,}000 + 150{,}000/(1.1) + 200{,}000/(1.1)^{\wedge}2 + 250{,}000/(1.1)^{\wedge}3 + 150{,}000/(1.1)^{\wedge}4 + 100{,}000/(1.1)^{\wedge}5 = 154{,}025$$

$$\text{NPV (B)} = -300{,}000 + 50{,}000/(1.1) + 150{,}000/(1.1)^{\wedge}2 + 200{,}000/(1.1)^{\wedge}3 + 300{,}000/(1.1)^{\wedge}4 + 200{,}000/(1.1)^{\wedge}5 = 348{,}772$$

The company should go for Proposal B.

B. *Internal Rate of Return*

Rate of return is generally defined as the earnings realized by a project in percentage of its principal capital.

The *IRR* is the average rate of return (usually annual) realized by a project in which the total net cash in-flow is exactly balanced with its total net cash out-flow, resulting in zero NPV value at the end of its project life cycle. In other words, this is the rate realizable when reinvestment of the project earnings is made at the same rate until maturity.

IRR is determined by the following equations:

$$0 = -P + \sum \frac{NCIF(m)}{(1 + IRR)^{\wedge} m} + \frac{CR}{(1 + IRR)^{\wedge} n} \tag{2.7}$$

$m = 1$ to n

For NCIF(m) = CF = constant

$$0 = -P + CF \frac{(1 + IRR)^{\wedge} n - 1}{IRR \times (1 + IRR)^{\wedge} n} + \frac{CR}{(1 + IRR)^{\wedge} n} \tag{2.8}$$

The IRR values (before tax) of acceptable projects must be greater than the firm's cost of capital. Projects with high IRR are to be preferred.

C. *Payback Period*

The payback period (PB) is defined as the number of years that the original capital investment for the project will take to be paid back by its annual earnings, or

$$\text{PB} = P/\text{CF} \tag{2.9}$$

where:

P = capital investment
CF = annual cash flow realized by the project

Cost reduction projects with small payback periods (e.g., less than 2 years) are preferable.

D. *Profitability Index*

Profitability index is defined by the ratio

$$PI = \frac{\text{Present value of all future benefits}}{\text{Initial Investment}} \tag{2.10}$$

Projects with large PI values are preferable.

Example 2.7

Your company is currently pursuing three cost-reduction projects at the same time.

1. Project A requires an investment of $10 million. It is expected to yield a cost saving of $30 million in the first year and another $10 million in the second year.
2. Project B demands an investment of $5 million. It is expected to produce a cost saving of $5 million in the first year and another $20 million in the second year.
3. Project C needs an investment of $5 million. It is expected to bring about a cost saving of $5 million in the first year and another $15 million in the second year.

After the second year, there will be no receivable benefit or capital recovery from any of these projects. The cost of capital (interest rate) is 10 percent.

Determine the ranking of these projects on the basis of the evaluation criteria of NPV, IRR, PB, and PI.

Answer 2.7

Let us define:

P = variable

$n = 2$

CF = variable

CR = 0

i = 10%

NPV:

A. NPV = –10 + 30/(1.1) + 10/(1.1)^2 = 25.537
B. NPV = –5 + 5/1.1 + 20/(1.1)^2 = 16.074
C. NPV = –5 + 5/1.1 + 15/(1.1)^2 = 11.942

IRR:

A. 0 = –10 + 30/(1 + *r*) + 10/(1 + *r*)^2; *r* = 2.3
B. 0 = –5 + 5/(1 + *r*) + 20/(1 + *r*)^2; *r* = 1.56
C. 0 = –5 + 5/(1 + *r*) + 15/(1 + *r*)^2; *r* = 1.3

PB:

A. PB = 10/[(30 + 10)/2] = 10/20 = 0.5
B. PB = 5/[(5 + 20)/2] = 5/12.5 = 0.4
C. PB = 5/[(5 + 15)/2] = 5/10 = 0.5

Table 2.17 Summary of results

Project	Time ≥ 0	1	2	NPV	IRR (%)	PB	PI
A	−10	30	10	25.5	230	0.5	3.55
B	−5	5	20	16	156	0.4	4.22
C	−5	5	15	12	130	0.5	3.39

PI:

A. PI = [30/1.1 + 10/(1.1)^2]/10 = 35.537/10 = 3.553
B. PI = [5/1.1 + 20/(1.1)^2]/5 = 21.074/5 = 4.214
C. PI = [5/1.1 + 15/(1.1)^2]/5 = 16.942/5 = 3.3884

Table 2.17 summarizes the results.

2.9 CONCLUSION

This chapter reviews basic cost-accounting issues related to product and service costing. Product costs have direct and indirect cost components. While direct costs are relatively easy to assess, the indirect costs that account for overhead charges may need to be properly assessed by using tools such as ABC, which focuses on redistribution of overhead costs to various cost objects on a rational basis. While the application of the ABC technique does not in itself reduce cost or improve profitability, its outcome could be used as inputs to the initiation of new or improved corporate strategies (e.g., product pricing, customers acquisition, etc.), the successful implementation of which is likely to lead to improved corporate profitability.

Cost data may be uncertain because of factors related to changes in economy, market condition, political stability, labor movement, and others. For uncertain cost data, risk analysis may be needed. The Monte Carlo simulation is an efficacious method to conduct risk analyses. Several other methods are also available to account for cost uncertainties.

Reviewed briefly in Appendix are some basic concepts in engineering economy. Cost data may apply for a single period or for multiple periods. In the case of multiple periods, the time dependency of cost data must be considered. For time-dependent data, the concept of the time value of money and the compound interest formulas are to be applied.

Depreciation accounting affects the facility costs that are part of the indirect costs of products. Different depreciation methods will lead to more or less indirect costs for the products. Finally, the inventory costs are affected by the sequence in which the time-dependent product costs are selected.

Benjamin Franklin said, "By failing to prepare, you are preparing to fail." It is indeed important for engineering managers and professionals to prepare themselves adequately by becoming well versed in cost accounting.

2.10 APPENDICES

The following appendices review a few basic concepts and tools in engineering economies.

2.10.1 COST ANALYSIS

Managers perform variance analyses and study the reasons for the deviation of actual costs from standard costs. They issue periodic and systematic reports of their findings and take proper actions to improve the efficiency and effectiveness of the organizational units.[25]

The two major factors affecting cost analysis are time and accuracy. For management decisions, cost analyses may be performed for a single time period or for multiple periods. Cost data may change over time or may be uncertain.

A. ***Single-Period Analysis***

Single-period analysis applies primarily to a short period of time during which the costs involved remain essentially constant. The gross profit (GP) equation for a given product line is given by the following equation:

$$\text{GP} = \text{revenue} - \text{costs}$$
$$\text{GP} = P \times N - (\text{FC} + \text{VC} \times N) \tag{2.11}$$

where P = product price (\$/unit)
N = number of products sold during the period (–)
FC = fixed costs (\$)
VC = variable costs (\$/unit)
GP = gross profit (\$)

For the case of break-even (i.e., GP = 0), the break-even product quantity is given by

$$N^* = \mathrm{FC}/(P - \mathrm{VC}) \quad (2.12)$$

The value (P – VC) is defined as the contribution margin of the product. Selling each additional unit of a product generates a contribution in the amount (P – VC) to defray the FC that has been committed to the production line.

Organizational performance can be readily assessed as the number of cost items involved is limited. One needs to make sure that the values of these cost items are valid, although from time to time the validity of such values may be tough to verify precisely, because of joint production activities and other cost-sharing systems involved.

B. *Multiple-Period Analyses*

The cost analyses over a longer period of time (e.g., multiple periods) are much more difficult to calculate for two reasons. First, costs may change predictably over time due to inflation, investment return, cost of capital, and other reasons. Second, future events are unpredictable (e.g., natural disasters, labor unrest, political instability, war against terrorism, spread of disease, investment climate, etc.).[26]

The change of costs over time needs to be addressed by using concepts such as NPV and IRR. These concepts are built on the fundamentals of the time value of money, compound interest, and the cost of capital. These topics are introduced in Appendix 2.10.2. Depreciation accounting, an important part of the indirect costs of products, is included in Appendix 2.10.3.

In dealing with the uncertainties of future costs, risks must be included in product cost analysis. Risk analysis has been elucidated in detail in Section 2.6.

2.10.2 TIME VALUE OF MONEY AND COMPOUND INTEREST EQUATIONS

The time value of money refers to the notion that the value of money changes with time. This is because money at hand may lose value (purchasing power) if not invested properly. Money at hand may earn income through investment. A dollar that is to be received at a future date is not worth as much as a dollar that is on hand at the present time. Thus, two accounts of equal dollar amounts but at different points in time do not have equal value (purchasing power).

Before introducing basic compound interest equations which are used for multiperiod cost analyses, a few definitions are reviewed next:

1. **Interest:** It represents a fraction of the principal designated as a reward (interest income) to its owner for having given up the right to use the principal. It may also be a charge (interest payment) to the borrower for having received the right to use the principal during a given interest period.
2. **Compound interest:** When the interest income earned in one interest period is added to the principal, the principal becomes larger for the next period. The enlarged principal earns additional interest under such circumstances. The interest is said to have been compounding.
3. **Nominal interest rate:** The interest rate quoted by banks or other lenders on an annual basis, also called the annual percentage rate (APR).
4. **Effective interest rate:** The interest rate in effect for a given interest period (e.g., one month). For example, if the nominal interest rate for a bank loan is 12 percent, then its effective interest rate for each month is 1 percent.
5. **Nominal dollar:** The actual dollar value at a given point in time.
6. **Constant dollar:** The dollar value that has a constant purchasing power with respect to a given base year (e.g., the reference year 2014); the value is adjusted for inflation.
7. **Consumer price index:** The index tracked by the U.S. Department of Commerce to indicate the price change for a basket of consumer products. Since 1993, the inflation rate in the United States has been relatively low.

To introduce the compound interest formulas for multiple-period cost analyses, the following notations are used:

P = present value ($): the value of a project, loan, or financial activity at the present time.

F = future value ($): the value of a project, loan, or financial activity at a future point in time.

i = effective interest rate for a given period during which the interest is to be compounded (e.g., 1 percent per month).

A = annuity ($): a series of payments made or received at the end of each interest period

n = number of interest periods (–).

A. Single Payment Compound Amount Factor

$$F = P \times (1 + i)^n \qquad (2.13)$$

$$F/P = (1 + i)^n = (F/P, i, n)$$

Equation 2.13 defines the total value of an investment P, with periodical returns added to the principals to earn more money at the end of n periods.

B. Present Worth Factor

$$P = F / (1 + i)^n \qquad (2.14)$$

$$P/F = (1 + i)^{-n} = (P/F, i, n)$$

Equation 2.14 defines the present value of a sum that will be available in the future. The factor is also called the discount factor.

C. Uniform Series Compound Amount Factor

$$F = A \times [(1 + i)^n - 1]/i \qquad (2.15)$$

$$F/A = [(1 + i)^n - 1]/i = (F/A, i, n)$$

Equation 2.15 determines the total future value of an account (e.g., retirement, college education, etc.) at the end of n periods, if a known annuity A is deposited into the account at the end of every period.

D. Uniform Series Sinking Fund Factor

$$A = F \times i/[(1 + i)^n - 1] \qquad (2.16)$$

$$A/F = i/[(1 + i)^n - 1] = (A/F, i, n)$$

Equation 2.16 calculates the amount of the required annuity (e.g., a series of period-end payments) that must be periodically deposited into an account in order to reach a desired total future sum F at the end of n periods.

E. Uniform Series Capital Recovery Factor

$$A = P\frac{i \times (1+i)^n}{(1+i)^n - 1} \qquad (2.17)$$

$$A/P = \frac{i \times (1+i)^{\wedge} n}{(1+i)^{\wedge} n - 1} = (A/P, i, n)$$

Equation 2.17 defines the amount of periodical withdrawal that can be made over n periods from an account worth P at the present time, such that the account will be completely depleted at the end of n periods.

F. Uniform Series Present Worth Factor

$$P = A\frac{(1+i)^{\wedge} n - 1}{i \times (1+i)^{\wedge} n} \quad (2.18)$$

$$P/A = \frac{(1+i)^{\wedge} n - 1}{i \times (1+i)^{\wedge} n} = (P/A, i, n)$$

Equation 2.18 determines the total present value of an account to which an annuity A is deposited at the end of each period. For example, if A is the periodical maintenance costs for capital equipment, then this equation calculates the present value of all maintenance costs over its product life of n periods.

Example 2.8

You need a new Apple MacBook Pro Laptop (15.4 Inch) with Retina Display, including 2.3 GHz Intel Core i7, 16 GB SDRAM, 512 GB Serial ATA and Mac OS *X* Marvericks. It is selling at $2,594.00, or $300 down and $120 monthly payment for 2 years. What nominal annual rate of return does this time-payment plan represent?

Answer 2.8

Loan amount = $2,594 – 300 = $2,294 = P

N = 24 months

A = $100

$P = A\ [(1 + i)^\wedge N - 1]/[i\ (1 + i)^\wedge N]$

$2{,}294 = 120\ [(1 + i)^\wedge 24 - 1]/[i\ (1 + i)^\wedge 24]$

This is an implicit equation for the single unknown i. It may be solved by trial and error: Input a trial value of i, and compute the value of the right hand side. The i value that balances the above equation is the answer.

Trial value of *i*	RHS
0.04	1,829.63
0.02	2,269.2
0.01	2,549.16
0.019	2,295.58
0.01908	2,293.50

Answer: Monthly rate = 1.908 percent and the nominal annual rate is 22.9 percent.

For all multiple-period problems, the timeline convention shown in Figure 1.23 is regarded as standard. When applying the above-described compound interest formulas, the following convention should be kept in mind:

1. **P** is at present, **F** is at a future point in time, and **A** occurs at the end of each period.
2. The periods must be consecutively and sequentially linked with the end of one as the beginning of the next.
3. Complex problems may be broken down into time segments so that the equations may be correctly applied to each of the segments.

Example 2.9

Company *X* manufactures automotive door panels that may be made of either sheet metal or plastic sheet molding (glass–fiber-reinforced polymer). Sheet metal bends well to the high-volume stamping process and has a low material cost. Plastic sheet molding meets the required strength and corrosion resistance and has a lower weight. The plastic-forming process involves a chemical reaction and has a slower cycle time. Table 2.18 summarizes the cost components for each.

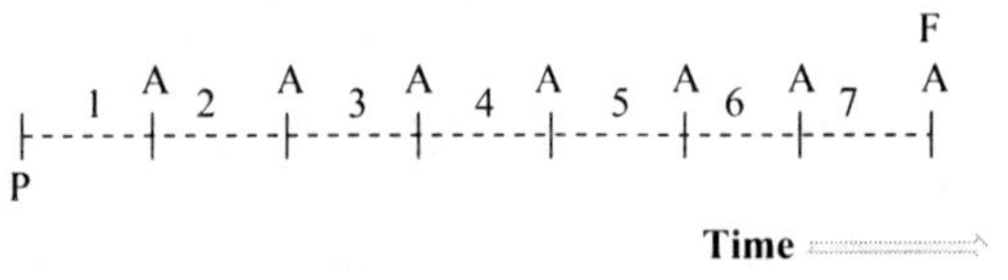

```
                                                F
      1  A  2   A  3  A  4   A  5   A  6  A  7  A
|-----|-----|-----|-----|-----|-----|-----|
P
                                  Time -------->
```

Figure 2.13. Timeline convention.

Table 2.18. Materials options

Description	Plastic	Sheet metal
Material cost (dollars per panel)	5	2
Direct labor cost (dollars per hour)	40	40
Factory overhead (dollars per year)	500,000	400,000
Maintenance expenses (dollars per year)	100,000	80,000
Machinery investment (dollars)	3,000,000	25,000,000
Tooling investment (dollars)	1,000,000	4,000,000
Equipment life (years)	10	15
Cycle time (minutes per panel)	2	0.1
Interest rate (percent)	6	6

Assuming that the machinery and tooling have no salvage value at the end of their respective equipment lives, what is the annual production volume that would make the plastic panel more economical?

Answer 2.9

This problem can be solved using the AC equation (Equation 2.2). Assume X is annual production volume, then we have the following results:

A. Plastic Panels

$$AC_1 = (3{,}000{,}000 + 1{,}000{,}000)\ (0.06)/\ [1.06\text{^}10 - 1] + 4{,}000{,}000 \times 0.06 + (500{,}000 + 100{,}000) + [5 + 40 \times 2/60] \times X$$

B. Steel Panels

$$AC_2 = [(25{,}000{,}000 + 4{,}000{,}000)\ (0.06)/[1.06\text{^}15 - 1)] + 29{,}000{,}000 \times 0.06 + (400{,}000 + 80{,}000) + [2 + 40 \times 0.1/60]\ 8X$$

Let $AC_1 = AC_2$; $X = 544{,}322$

For production volume up to 544,322 panels per year, the plastic panels are more economical.

2.10.3 Depreciation Accounting

In calculating indirect costs associated with production facility, equipment, and other tangible assets related to production, depreciation charges must be included. Depreciation is a cost-allocation procedure whereby the cost of a long-lived asset is recognized in each accounting period over the asset's useful life in proportion to its usage brought forth over the same period. This procedure is undertaken in a reasonable and orderly fashion. Specifically, the acquisition cost of an asset can be considered as the price paid for a series of future benefits. As the asset is partially used up in each accounting period, a corresponding portion of the original investment in the asset is treated as the cost incurred for the partial benefit delivered.

The U.S. Internal Revenue Service accepts three depreciation accounting methods. They are discussed next, using the following notations:

P = initial investment ($) at the present time.

N = useful life of a long-lived asset measured in years (e.g., $N = 25$ for buildings, $N = 15$ for equipment, $N = 5$ for automobiles, $N = 3$ for computers, etc.).

$D(m)$ = depreciation charge ($) in the asset's m(th) year.

L = salvage value ($) recoverable at the end of the equipment's useful life.

$\text{AD}(m)$ = accumulated depreciation ($), which is the total amount of depreciation charges accumulated at the end of the m(th) year.

$\text{BV}(m)$ = book value ($) of an asset in its m(th) year; $\text{BV}(m) = P - \text{AD}(m)$.

$P - L$ = depreciable base ($).

$r(m)$ = depreciation rate, a fraction of the depreciable base to be depreciated per year.

A. *Straight Line*

By this depreciation method, an equal portion of depreciation base $(P - L)$ is designated as the depreciation charge for each period of the assets' estimated useful life:

$$D(m) = (P - L)/N = \text{constant} \tag{2.19}$$

$$\text{BV}(m) = P - m \times (P - L)/N;\ m = 1,2,3,\ldots$$
$$r(m) = 1/N = \text{constant}$$
$$\text{AD}(m) = m \times (P - L)/N$$

More than 91 percent of publicly traded companies in the United States apply this straight-line depreciation method.

B. *Declining Balance*

The depreciation charge is set to equal to the net book value (e.g., acquisition cost minus accumulated depreciation) at the beginning of each period (e.g., year) multiplied by a fixed percentage. If this percentage is two times the straight-line depreciation percentage, then it is called a double-declining balance method:

$$D(m) = P \times r \times (1 - r)^{\wedge}(m - 1) \quad (2.20)$$

$BV(m) = P \times (1 - r)^{\wedge}m$

$r(m)$ = constant; $r = 2/N$ (double-declining balance method)

$AD(m) = P \times [1 - (1 - r)^{\wedge}m]$

Note that the salvage value is not subtracted from the acquisition cost. To make sure that the total accumulated depreciation does not exceed the depreciation base, the depreciation charge of the very last period (e.g., year) must be manually adjusted.

C. *Units of Production Method*

The depreciation charge is assumed to be proportional to the service performed (e.g., units produced, hours consumed, etc.). Companies that are involved with natural resources (e.g., oil and gas exploration) use units of production method to depreciate their production assets. Software companies also use this method to depreciate their capitalized software development costs.

Example 2.10

The company plans to change its depreciation accounting from the straight-line method to the double-declining method on a class of assets that have a first cost (acquisition cost) of $80,000, an expected life of six years, and no salvage value. If the company's tax rate is 50 percent, what is the present value of this change, assuming 10 percent interest compounded annually?

Answer 2.10

$P = 80{,}000$; $N = 6$; $t = 0.5$; $L = 0$

$F = (1 - t)$ Delta; $P = F/(1 + i)^{\wedge}n$; $i = 10\%$

Table 2.19 shows the present values of the differences between these two depreciation charges for the assets' expected life of six years.

Table 2.19. Calculation of difference due to depreciation methods

Year	Straight line	Double-declining ($r = 2\ (1/6) =$ 0.333333)	Delta	$F = (1 - t) \times$ Delta	Present worth
1	13,333.33	26,666.67	13,333.34	6,666.67	6,060.61
2	13,333.33	17,778.66	4,445.33	2,222.66	1,836.91
3	13,333.33	11,851.87	−1,481.46	−740.73	−556.52
4	13,333.33	7,901.25	−5,432.08	−2,716.04	−1,855.09
5	13,333.33	5,267.5	−8,065.83	−4,032.92	−2,504.12
6	13,333.33	10,534.05	−2,799.28	−1,399.64	−790.06
			Total		2,191.72

Answer = $2,191.72

Example 2.11

A new delivery truck costs $40,000 and is to be operated at approximately the same amount each year. If annual maintenance costs are $1,000 the first year and increase $1,000 each succeeding year and, if the truck trade-in value is $24,000 the first year and decreases uniformly by $3,000 each year thereafter, at the end of which year will the costs per year of ownership and maintenance be a minimum?

Answer 2.11

The average annual ownership cost is calculated as shown in Table 2.20.

It shows that the combined ownership and maintenance cost reaches a minimum in 5th year.

2.10.4 INVENTORY ACCOUNTING

After the direct and indirect costs are estimated, the product costs can be defined. When products are transferred from WIP operations to a FG warehouse, they become inventory. Inventory may be managed by one of two methods: first in and first out (FIFO) and last in and first out (LIFO). The FIFO method specifies that inventory that enters the warehouse first will leave the warehouse first. By the LIFO method, the inventory that enters the warehouse last is shipped out first.

According to the time value of money concept, these two inventory operational methods may render different CGS. The inventory accounting takes into account such possible change of product cost over time, due to, for example, inflation. In general, companies utilize one of the following three inventory accounting methods:

A. FIFO
B. LIFO
C. Weighted average

LIFO is most useful during periods of high inflation, as it results in less reportable earnings with lower taxes paid; LIFO is not useful, however, when prices for raw materials decrease. LIFO also provides a lower inventory value, thus understating the value of the inventory in the balance sheet. Finally, LIFO is a more conservative accounting technique than

Table 2.20. Annual ownership cost computation

Year	Maintenance cost (annual)	Trade-in value	Accumulated depreciation	Total accumulated cost of ownership	Average annual cost over the ownership period
1	1,000	24,000	16000	17,000	17,000
2	2,000	21,000	19,000	22,000	11,000
3	3,000	18,000	22,000	28,000	9,333
4	4,000	15,000	25,000	35,000	8,750
5	5,000	12,000	28,000	43,000	8,600
6	6,000	9,000	31,000	52,000	8.667

FIFO. Note that LIFO is prohibited by law in some countries, such as the United Kingdom, France, and Australia.

As a product of creative accounting, FIFO defines an inventory value more closely matched with its market value. It tends to make the income statement look better than it really is. In periods when the business climate experiences stagnation or recession, innumerable companies frequently switch from LIFO to FIFO. The weighted average method represents a compromise between the two.

Table 2.21 is an illustration of the use of FIFO and LIFO accounting techniques. Assume that a manufacturing company has five units of products in inventory and each has a product cost of $100. Furthermore, the company produces five more units at $200 each in one period and then another five units at $300 each in a later period. During these periods, the company sells 10 units to customers. Determine the average CGS on the basis of both FIFO and LIFO and assess its impact on the company's net income.

Table 2.21. Inventory accounting

S. No.	Chronology of transactions	FIFO ($)	LIFO ($)	Average ($)	Inventory Actions
1	Beginning inventory				
	5 × $100	500	500	500	Withdrawal of 10 units
					—
2	Purchase and value added				5 × 100
	5 × $200	1,000	1,000	1,000	5 × 200
					5 × 300
	5 × $300	1,500	1,500	1,500	—
3	Ending inventory				
	5 *	1,500	500	1,000	
4	Cost of goods sold	1,500	2,500	2,000	

Table 2.22. Effect of inventory accounting on net income

Income statement entries	FIFO ($)	LIFO ($)	Weighted average ($)
Sales	10,000	10,000	10,000
CGS	*1,500*	*2,500*	*2,000*
Gross margin	8,500	7,500	8,000
GS&A	*2,000*	*2,000*	*2,000*
EBIT	6,500	5,500	6,000
Interest	500	500	500
Taxable amount	6,000	5,000	5,500
Tax (40%)	*2,400*	*2,000*	*2,200*
Net Income	3,600	3,000	3,300

The impact of inventory accounting on net income is quite direct, as illustrated in Table 2.22, an abbreviated income statement, where CGS is costs of goods sold, GS&A is general, sales, and administration expenses, and EBIT is earnings before interests and taxes. Upon switching from FIFO to LIFO inventory accounting, the tax liabilities are shown to have been reduced form $2.4 million to $2.0 million.

2.10.5 CONVERSION OF A PROBABILITY DENSITY FUNCTION TO ITS CUMULATIVE DISTRIBUTION FUNCTION

The process of converting a probability density function to its cumulative distribution function is straightforward and unique. Figure 2.14 shows a triangular probability density function for the cost of the component C1. The vertical axis represents probability, and the horizontal axis represents cost. The triangular probability density function is the easiest one to apply when a three-point estimate for a risky input variable is known.

The component C1 is assumed to have a minimum cost of $30 (point *A*), a maximum cost of $80 (point *Z*), and a most likely cost of $50 (point *M*). The area underneath the triangular probability density function is normalized to 1. This condition prescribes that the y coordinate for the point N is 0.04 based on the calculation of $1 = 0.04 \times 0.5 \times (80 - 30)$.

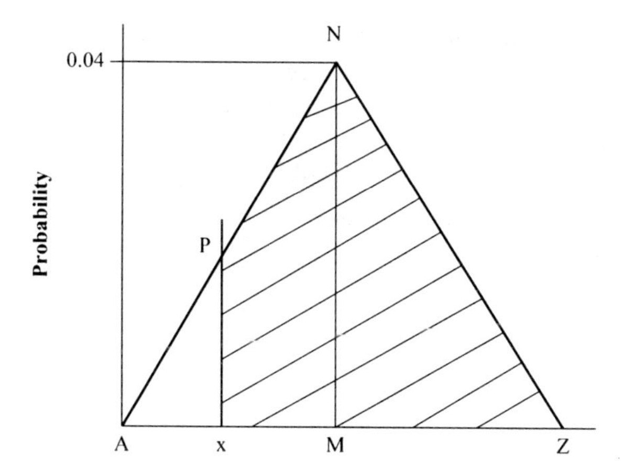

Figure 2.14. Conversion of a triangular probability density function to its cumulative descending function.

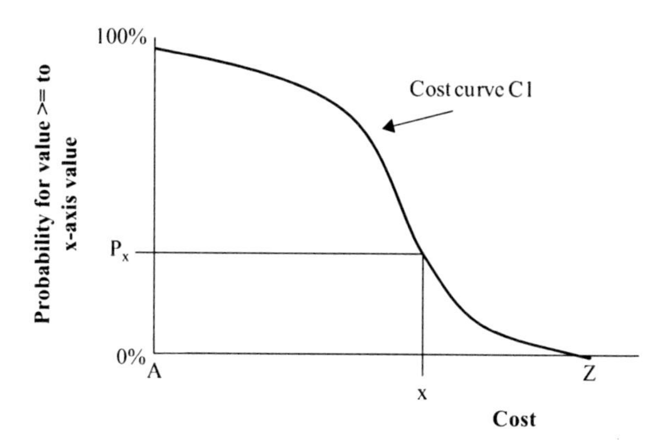

Figure 2.15. Cumulative descending distribution function of a cost component.

Let us insert a vertical cost line through x. With this cost line in place, we define the shaded area PNZXP as A_x, which is under the probability density function, but bound by the vertical cost line that passes through x on the left. We form a ratio of A_x to the total area APNZXA underneath the same probability density function. This ratio is designated as P_x. The value of P_x varies from 0 to 1, as x moves from Z to M and then to A. P_x is the probability for the cost of this component to be equal to or in excess of x. The pair of P_x and x represents a point in a cumulative distribution chart (Figure 2.15).

For another cost value, y, this process is repeated. A new pair of Py and y defines another point in the descending cumulative distribution chart. After many repetitions, a descending cumulative curve is generated that resembles the one shown in Figure 2.15. The vertical axis is the probability for the component cost to equal or exceed the value shown on the x axis. The x axis spans the minimum value of A on the left to the maximum value of Z on the right.

CHAPTER 3

FINANCIAL ACCOUNTING AND ANALYSIS

3.1 INTRODUCTION

Financial accounting and analysis serve the important corporate functions of reporting and evaluating the financial health of a firm.[1] Financial statements are prepared by certified management accountants (CMAs) and certified public accountants (CPAs), according to the Generally Accepted Accounting Principles (GAAP) in a conservative, material, and consistent manner. These financial documents provide (a) internal reporting to corporate insiders for planning and controlling routine operations and for decisions on capital investments, and (b) external reporting to shareholders and potential investors in financial markets.[2] The financial performance of a company always garners a lot of attention.

All financial statements are designed to be relevant, reliable, comparable, and consistent. Financial accounting treats owners (shareholders) and corporations as separate entities. Owners of corporations are liable only to the extent of their committed investments. Owners enjoy a flexible tenure and participation, as they may buy or sell stocks of the company at any time. On the other hand, corporations are legal entities, fully responsible for their liabilities up to the limits of their total assets. All corporations are assumed to be going concerns and in operation forever, unless they cease to exist by declaring bankruptcy or being acquired by others.

Managerial finance focuses on the sources and uses of funds in a company. Capital budgeting is the key responsibility of the company's top management. Capital budgeting decisions take into account the need to expand production facilities, develop new products, create new supply chains, acquire new technologies to complement the company's own core competencies, penetrate into new regional markets, and indulge in other corporate activities. Besides defining what is worth doing, top management must also find the investment capital needed to execute the chosen short-term and long-term strategies.[3]

Resource allocation drives the company's overall performance.[4] Allocation decisions address three major ways of utilizing investment capital: (a) assets in place—building new facilities and developing new products, (b) marketing and R&D efforts, and (c) strategic partnerships—acquisitions and joint ventures. In deciding on resource allocation, company management needs to know what a specific endeavor is worth. Thus, different ways of estimating the value of companies, projects, and opportunities need to be established.

To be covered in this chapter are (a) language and concepts; (b) financial statements; (c) performance ratios and analysis; (d) balanced scorecards or tools to monitor and promote corporate productivity; (e) the valuation of assets in place by the discounted cash flow (DCF) methods, and R&D and marketing opportunities by the simple option method; and (f) the use of T-accounts. No discussions about the valuation of strategic partnerships and capital formation via equity and debt financing are included, as engineering managers and pofessionals are not often involved in the detailed financing aspects of raising capital for acquisitions, and joint ventures.

Engineering managers and professionals need to know how to read financial statements; monitor the firm's activity, performance, profitability, and market position; and assess the long-term health of a firm. They also need to understand the language, principles, and practices of financial management.[5] Doing so will allow them to bring clarity when initiating proper projects (e.g., plant expansion, new product and technology development, new technology acquisition, strategic alliances, etc.) at the right time to add value to their employers as well as to participate in decision making that involves major projects in which engineering managers provide significant inputs.

3.2 FINANCIAL ACCOUNTING PRINCIPLES

As practiced in the United States, all financial statements are formulated for a specific accounting period. A typical accounting period is three months, as the U.S. Securities and Exchange Commission prescribes that all publicly traded companies file Form 10-Q reports every quarter. All companies also need to publish their Form 10-K reports annually. The financial statements must adhere to the basic principles of accounting, discussed in the following subsections.[6]

3.2.1 ACCRUAL PRINCIPLE

Accounting statements include both cash and credit transactions. Revenue is recognized when it is earned. For example, a manufacturing enterprise will recognize revenues as soon as products are shipped to the customer and an invoice is sent, irrespective of any credit payment already received or yet to be collected. Sports teams are known to sell season tickets ahead of the games for cash and then recognize the applicable revenue only after each game is played. According to the accrual principle of accounting, companies recognize revenues when earned, with the assumption that the collection of this revenue from approved credit accounts and the delivery of the promised products or services are both reasonably assured.

Similarly, the accrual principle specifies that costs and expenses are established when incurred, even before actual payments are made.

3.2.2 MATCHING

Expenses are recognized by matching them with the revenue generated in a given accounting period. For example, the cost of goods sold (CGS) is recognized as an expense only after products are sold and revenue is recognized. Before the products are sold, CGS stays as inventory—a part of the corporate current assets—even though costs for materials, labor, and factory overhead have already been spent for these yet-to-be-sold products.

3.2.3 DUAL ASPECTS

The assets of a company are always equal to the claims against them (i.e., assets equal to claims). The claims originate from both creditors and owners. Each transaction has a dual effect in that it induces two entries in order to maintain a balance between assets and claims; see T-Accounts (Appendix 3.8.1).

3.2.4 FULL DISCLOSURE PRINCIPLE

All relevant information is disclosed to the users of the company's financial reports. Extensive footnotes contained in the annual reports of numerous publicly traded companies are testimonials for such disclosure practices.

3.2.5 CONSERVATISM

Assets are to be recorded at the lowest value consistent with objectivity (e.g., book values are often lower than market values). While profits are not recorded until recognized, losses are recorded as soon as they become known. Inventories are valued at the lower of the cost or market value.

3.2.6 GOING CONCERN

It is generally assumed that the company's business will go on forever. This assumption justifies the current practice of using historical data (e.g., the original acquisition costs) and a reasonable method of depreciation (e.g., straight line) by which the book value of corporate tangible assets is defined. Otherwise, liquidation accounting must be applied to define the corporate asset value by using current market prices.[7]

3.3 KEY FINANCIAL STATEMENTS

Companies in the United States use three financial statements: income statement, balance sheet, and funds flow statement.[8]

3.3.1 INCOME STATEMENT

The income statement is an accounting report that matches sales revenue with pertinent expenses that have been incurred (e.g., CGS, taxes, interest, depreciation charges, salaries and wages, administrative expenses, R&D, etc.). Sometimes it is also called the profit or loss statement, earnings statement, or operating and revenue statement. The income statement contains the following key entries:

1. *Sales revenue* is the total revenue realized by the firm during an accounting period. Sales revenue is recognized when earned, for example, by having goods shipped and invoices issued.
2. *CGS* is the cost of goods that have been actually sold during an accounting period. In a manufacturing company, CGS is calculated as the opening inventory at the beginning of an accounting period, plus labor costs, material costs, and manufacturing overhead incurred during the period, and minus the closing inventory at the end of the period.

3. *Gross margin* is the sales revenue minus the CGS. The gross margin percentage is defined as the ratio of gross margin to sales revenue. This percentage is typically in the range of 5 percent to 40 percent.
4. *Expenses* are those expenditures chargeable against sales revenue during an accounting period. Examples include general, selling, and administrative expenses; depreciation charges; R&D expenses; advertising expenses; interest payments for bonds; employee retirement benefit payments; and local taxes.
5. *Depreciation* is a process by which the cost of a fixed, long-lived asset is converted into expenses over its useful life. This noncash exependiture is to be claimed in proportion to the value it has produced during an accounting period (see Appendix 2.10.3).
6. *Earning before interest and taxes (EBIT)* is the earnings before interests and taxes. Sometimes, it is also called operational income.
7. *Net income (earnings or net operating profit after tax [NOPAT])* is the excess of sales revenue over all expenses (e.g., CGS, all items under 4 above, and corporate tax) in an accounting period. Sometimes it is also called profit or earnings.
8. *Dividend* is the amount per share paid out to stockholders in an accounting period. Typically, the company's board of directors declares the percentage of the net income (say 30 percent) earned in a given period to be paid out to shareholders as dividend. The remainder portion (say 70 percent) of the net income goes into the retained earning account, to be used for investment purposes in the subsequent periods by the company manangement.
9. *Earnings per share* (EPS) are the net income of a firm during an accounting period (e.g., a year), minus dividends on preferred stock, divided by the number of common shares outstanding.
10. *Costs* are resources in dollar spent by the company in a given accounting period. Costs differ from expenses. Expenses are those costs that are chargeable against revenues in a given accounting period. For example, direct and indirect costs contained in the products preserved as inventory are not recognized immediately as expenses. When products in inventory are sold, then the respective CGS is recognized as expenses in the income statement, along with other expenses.
11. *Cash flow* is the sum of net income earned plus the depreciation charge claimed in a given accounting period.

An income statement shows the firm's activity. An example is given in Table 3.1 for the XYZ Company. In general, sales revenue is referred to as

Table 3.1. Example of XYZ income statement

Entries	Year 2011 (millions of dollars)	Year 2012 (millions of dollars)
Sales (net) revenue	8,380.30	8,724.70
Cost of goods sold	6,181.20	6,728.80
Gross margin	2,199.10	1,995.90
General, selling, and administrative expenses	320.70	318.80
Pensions, benefits, R&D, insurance, and others	494.60	538.70
State, local, and miscellaneous taxes	180.10	197.10
Depreciation	297.20	308.60
EBIT	906.50	632.70
Interest and other costs related to debts	82.90	114.40
Corporate tax	(32.05%) 264.00	(20.84%) 108.00
Net income (NOPAT)	559.60	410.30
Common stock dividend	151.60	172.80
Retained earnings	408.00	237.50

the *top line* and net income as the *bottom line* figures in an income statement. These line items are examined closely by financial analysts, as are the line items of gross margin and EBIT. (For a detailed analysis and interpretation of income statement entries, see Section 3.4.2 on Ratio Analysis.)

Engineering managers and professionals are involved in deploying company resources to promote the financial success of their employers. The impact of their engineering activities is registered in several line items contained in the income statement:

1. **Sales:** Engineering managers increase sales through well-designed products that satisfy the needs of customers. They introduce innovative features and functionalities of their products/servcies that enhance their relative competitiveness in the marketplace. They refine products/services that are easy to serve and maintain, thus promoting their market acceptance. They also identify and secure suitable supply partners to increase the speed of introducing prod-

uct/service and the extent of product/service customization in the marketplace.

2. **CGS:** Engineering managers and professionals cut down product/service costs by performing work related to innovative design, value engineering, manufacturing, and quality control.
3. **R&D:** Engineering managers advance and apply new technologies to enhance product/service features and to foster the rapid development of new global offerings.

Example 3.1

The Advanced Technologies company has had quite a successful year. Its assets, liabilities, revenues, and expenses at the end of the current fiscal year are exhibited in Table 3.2.

Determine the net income of the company for the current year.

Answer 3.1

To determine the company's net income, we need to create the income statement for the company. As presented in Table 3.3, only selected items of Table 3.2 are to be included in the company's income statement. For the current year, the company has a net income of $2,050,000.

3.3.2 BALANCE SHEET

The balance sheet is an accounting report that presents the assets owned by a company and the ways in which these assets are financed through liabilities and owners' equity (OE). Equation 3.1 depicts the balance between assests (A) and claims consisting of liabilities (L) and OE.

$$A = L + \text{OE} \tag{3.1}$$

Liabilities are creditors' (such as banks, bondholders, and suppliers) claims against the company. Owners' equity represents the claims of owners (shareholders) against the company.[9] The following key entries are included in a balance sheet:

1. *Assets* are items of value having a measurable worth. They are resources of economic value possessed by the company. There are three classes of assets: current, fixed, and all others.

Table 3.2. Records of financial entries

#	Items	Thousands of dollars
1	Accounts payable	3,740
2	Accounts receivable	7,550
3	Advertising expense	3,340
4	Administrative expense	5,500
5	Building (net)	36,300
6	Cash	6,320
7	Cost of goods sold	31,000
8	Depreciation expense—building	960
9	Depreciation expense—equipment	1,310
10	Equipment (net)	14,640
11	Inventory	11,000
12	Insurance expense	840
13	Interest expense	2,100
14	Land	2,100
15	Long-term loans outstanding	42,000
16	Miscellaneous expense	1,480
17	R&D	5,200
18	Salaries payable	170
19	Sales revenue	60,300
20	Supplies expense	1,820
21	Taxes expense	2,630
22	Taxes payable	610
23	Utilities expense	2,070

2. *Current assets* are convertible to cash within 12 months. Examples include, in a descending order of liquidity, cash, marketable securities, accounts receivable, inventory, and prepaid expenses.
3. *Cash* is money on hand or in checks, and is the most liquid form of assets.

Table 3.3. Income statement of Advanced Technologies

Entries	Thousands of dollars
Sales revenue	60,300
Cost of goods sold	31,000
Gross margin	29,300
Administrative expense	5,500
Advertising expense	3,340
Supplies expense	1,820
Utilities expense	2,070
Miscellaneous expense	1,480
Insurance expense	840
Depreciation—building	960
Depreciation—equipment	1,310
R&D	5,200
Operating income	6,780
Interest expense	2,100
Taxable income	4,680
Taxes expense	2,630
Net income	2,050

4. *Accounts receivable* is the category of revenue recognized prior to payment collection. It is money owed to the firm, usually by its customers or debtors, as a result of a credit transaction.
5. *Inventory* designates stock of goods yet to be sold that is valued at cost, including direct materials, direct labor, and manufacturing overhead. It may consist of stores, work in progress, and finished goods inventories (see Appendix 2.10.4). Inventory is included in the balance sheet as a current asset.

   When finished goods are shipped and invoiced to customers in an accounting period, their respective CGS is then recognized in the income statement as an expense.
6. *Prepaid expenses* are paid before receiving the expected benefit (e.g., rent, journal subscription fee, or season's tickets). It is a current asset.

7. *Fixed assets* are tangible assets of long, useful life (more than 12 months) such as land, buildings, machines, and equipment. Improvement costs for these fixed assets are added to the fixed asset value. On the other hand, repair and maintenance costs for them incurred in a given accounting period are expensed in the income statement.
8. *Other assets* are valuable assets that are neither current nor fixed. Examples include patents, leases, franchises, copyrights, and goodwill. Amortization accounting applies to these assets in a similar manner, as depreciation is applied to fixed assets (see Appendix 2.10.3).

   Goodwill—a company's reputation and brand-name recognition—is recognized as an asset only if it has been purchased for a measurable monetary value, such as in conjunction with a merger or acquisition transaction.
9. *Accumulated depreciation* is the sum of all annual depreciation charges taken from the date at which the fixed asset is first deployed up to the present.
10. *Net fixed assets* are the net value of the firm's tangible assets—original acquisition cost minus accumulated depreciation. Note that this net fixed asset value may deviate considerably from its market value or replacement cost in a given accounting period. The conservatism principle prescribes that the net fixed asset is carried on the balance sheet even if it is lower than its current market value. Otherwise, a loss entry must be added to the balance sheet to adjust the fixed asset value downwards, should the net fixed asset become higher than its current market value.
11. *Liabilities* are obligations that are to be discharged by the company in the future. They represent claims of creditors (e.g., banks, bondholders, and suppliers) against the firm's assets. Sometimes, it is also called *debt*.
12. *Current liability* describes amounts due for payment within 12 months. Examples include accounts payable, short-term bank loans, interest payments, payable tax, insurance premiums, deferred income, and accrued expenses. Accounts payable is always listed first within the category of current liabilities, with others to follow in no specific order.
13. *Accounts payable* is an expense recognized before payment. It is an obligation to pay a creditor or a supplier as a result of a credit transaction, usually within a period of one to three months.
14. *Deferred income* is income received in advance of being earned and recognized (i.e., payment received before shipment or invoicing,

or both, of goods). In the balance sheet, it is included as a current liability.

15. *Deferred income tax* is the amount of tax due to be paid in the future, usually within 12 months.
16. *Long-term liability* is defined as the amounts due to be paid beyond the next 12 months. Examples include corporate bonds, mortgage loans, long-term loans, lines of credit, long-term leases, and contracts.
17. *Bonds* are long term debt certificates secured by the assets of a company or a government. Bonds issued by a publicly held company are corporate bonds, and those issued by the U.S. federal government are Treasury bonds. In case of defaults, bondholders have the legal right to seize the assets, which the company has placed as collaterals for the bonds, for recovery.
18. *Debentures* are unsecured bonds issued by the firm. Typically, debentures carry a higher interest rate than bonds that are secured by collaterals.
19. *Convertible bonds* are those issued by a company and they are allowed to be converted into common stocks according to a set of predetermined conversion schemes.
20. *Owners' equity* is the shareholders' original investment plus accumulated retained earnings. It represents the residual value of the corporation owned by the shareholders after having deducted all liabilities from company assets. Sometimes it is also called *net worth.*
21. *Stock* is a certificate of ownership of shares in a company. Preferred stocks have a fixed rate of dividend that must be paid before dividends are distributed to holders of common stocks.
22. *Capital surplus* is the premium price per share above the par value of the stock. It includes the increase in the owner's equity above and beyond the difference between assets and liabilities reported in the company's balance sheet. It is part of the owners' equity.
23. *Retained earnings* are the accumulated earnings retained by the company for the purpose of reinvestment and which had not been paid out as dividends.
24. *Book value* is defined as the tangible assets (such as fixed assets) minus liabilities and the equity of preferred stocks. It is the share value of common stocks carried in the books.
25. *Stock price* is the market value of a firm's stock. It is influenced by the book value, earning per share, anticipated future earnings, perceived management quality, and environmental factors present in the marketplace.

The contributions of engineering managers affect only one line item in the balance sheet, namely, inventories. Inventories may be scaled down by applying superior production technologies, product design, and best practices of supply-chain management.

The organization of entries in a balance sheet follows the specific convention listed below:

1. Assets are listed before liabilities, which are then followed by OE.
2. Current assets and liabilities are enumerated ahead of noncurrent assets and liabilities, respectively.
3. Liquid assets are listed before all other assets with less liquidity.
4. The listing of current liabilities follows no specific order, except that accounts payable must always be listed first in this section.

Table 3.4 shows a sample balance sheet of XYZ Company.

Table 3.4. Example of XYZ balance sheet

Entries	**Year 2011 (millions of dollars)**	**Year 2012 (millions of dollars)**
Assets		
Cash	231.00	245.70
Marketable securities	450.80	314.90
Accounts receivable	807.10	843.50
Inventories	1,170.70	1,387.10
Total current assets	2,659.60	2,791.20
Fixed assets	11,070.40	11,897.70
Accumulated depreciation	6,410.70	6,618.50
Net fixed assets	4,659.70	5,279.20
Long-term receivables and other investments	574.80	735.20
Prepaid expenses	260.90	362.30
Total long-term assets	5,495.40	6,376.70
Total assets	**8,155.00**	**9,167.90**

(Continued)

Table 3.4. (*Continued*)

Entries	Year 2011 (millions of dollars)	Year 2012 (millions of dollars)
Liabilities		
Accounts payable	571.20	622.80
Notes payable	65.30	144.50
Accrued taxes	346.30	275.00
Payroll and benefits payable	433.70	544.30
Long-term debt due within a year	30.40	50.80
Total current liabilities	1,446.90	1,637.40
Long-term debt	1,542.50	1,959.90
Deferred tax on income	288.40	405.30
Deferred credits	27.00	36.30
Total long-term liabilities	1,857.90	2,401.50
Total liabilities	**3,304.80**	**4,038.90**
Common stock ($1.00 par value)	81.40	82.20
Capital surplus	1,549.10	1,589.60
Accumulated retained earnings	3,219.70	3,457.20
Total owner's equity	**4,850.20**	**5,129.00**
Total liabilities and owners' equity	**8,155.00**	**9,167.90**

Example 3.2

Using the data given in Table 3.2, construct the balance sheet of Advanced Technologies and determine the OE at the end of the current fiscal year.

Answer 3.2

The OE amount to $29,120,000 at year end (see Table 3.5).

3.3.3 FUNDS FLOW STATEMENT

The funds flow statement compares the firm's activities in two consecutive accounting periods from the standpoint of funds. It is an accounting

Table 3.5. Balance sheet of Advanced Technologies

Entries	**Thousands of dollars**
Cash	6,320
Accounts receivable	7,550
Inventory	11,000
Total current assets	24,870
Land	2,100
Equipment (net) ($14,640–1,310)	13,330
Building (net) ($36,300–960)	35,340
Total assets	75,640
Accounts payable	3,740
Taxes payable	610
Salaries payable	170
Long-term loans outstanding	42,000
Total liabilities	46,520
Owners' equities	29,120
Total liabilities and owners' equities	75,640

report that elucidates the major sources and uses of funds of the firm. It is sometimes also called *statement of changes in financial position* or the *statement of sources and uses of funds*.

The principle behind the funds flow analysis is rather simple. An increase in assets signifies a use of funds, such as buying a plant facility by paying cash or using credit. A decrease in assets indicates a source of funds, such as selling used equipment to receive cash for use in the future. An increase in liabilities produces a source of funds, such as borrowing money from a bank so that more cash is available for other purposes. A decrease of liabilities yields a use of funds, such as paying down a bank loan by using money from the company's cash reservoir. Table 3.6 presents an example of the funds flow statement of XYZ Company at the end of year 2012, based on the financial data presented in Table 3.1.

The funds flow statement shown in Table 3.6 is generated by applying the following procedure:

Table 3.6. Example of XYZ funds flow statement

Entries	2011–2012 (millions of dollars)	Percentage
Sources		
Increase in long-term debt	437.40	26.50
Net income	410.30	24.86
Depreciation*	308.60	18.70
Decrease in marketable securities	135.90	8.23
Increase in deferred taxes on income	116.90	7.08
Increase in payroll and benefits payable	110.60	6.70
Increase in notes payable	79.20	4.80
Increase in accounts payable	51.60	3.13
Total source of funds	**1,650.50**	100.00
Uses		
Increase in fixed assets and other investments	928.10	56.23
Increase in inventories	216.40	13.11
Dividend paid	172.80	10.47
Increase in prepaid expenses	101.40	6.14
Increase in long-term receivables	59.60	3.61
Decrease in accrued taxes	71.30	4.32
Increase in capital surplus	40.50	2.45
Increase in accounts receivable	36.40	2.21
Increase in cash	14.70	0.89
Increase in deferred credits	9.30	0.56
Total uses of funds	**1650.50**	100

*Depreciation is a noncash expenditure that must be added back here to denote a source of funds available to the firm.

A. Increase in plants and equipment

1. Increase in fixed assets (11897.7 – 11070.4) = 827.3
2. Increase in long-term receivables and other investments (735.2 – 574.8) = 160.4
3. As details are missing, we introduce the following reasonable assumption:
   (a) Long-term receivables = 59.6
   (b) Increase in other investment = 100.8
4. Total increase in fixed asset investment (827.3 + 100.8) = 928.1

B. Increase in long-term debt

1. Increase in long-term debt (1959.9 – 1542.5) = 417.4
2. Increase in long-term debt due within one year (50.8 – 30.4) = 20.4
3. Total (417.4 + 20.4) = 437.8

C. Increase in common stock and capital surplus

1. Increase in common stocks (82.2 – 81.4) = 0.8
2. Increase in capital surplus (1589.6 – 1549.1) = 40.5
3. Total = 41.3

Most other line items in the statement are directly verifiable. By examining the individual percentages of sources and uses of funds involved, as shown in the last column in Table 3.6, managers and investors can readily evaluate the extent to which fund flow is aligned with the stated corporate strategies.

3.3.4 LINKAGE BETWEEN STATEMENTS

The three financial statements described previously are linked to one another. The net profit in the income statement is linked with the retained earning in the balance sheet. The inventory account in the balance sheet is linked with the sales revenue in the income statement. The accumulated depreciation in the balance sheet is linked with the annual depreciation charge included in the income statement. Because the depreciation charge taken in a given period affects the net profit of the company during the same period, it is thus indirectly linked to the retained earning account in the balance sheet as well.

The linkage between the funds flow statement and the other two financial statements is self-evident, as all data in the funds flow statement are derived from changes in various line items contained in the other two statements.

3.3.5 *RECOGNITION OF KEY ACCOUNTING ENTRIES*

This section offers additional notes on the recognition of several key accounting entries—assets, liabilities, revenues, and expenses—according to GAAP practiced in the United States. Other countries may have slightly different rules governing the reporting practices of these items.

A. ***Assets*** are the resources under company control. They have economic value and can be used to produce future benefits. Asset recognition is based on two principles: historical cost and conservatism.[10] All assets are reported by using historical cost—that is, the initial capital investment value at some time in the past. The book value of a given asset is defined as its initial acquisition cost minus the accumulated depreciation. Should the asset's current market value drop below its book value, the shortfall must be reported as an expense. If its market value exceeds its book value, however, the surplus is not reported in the company's balance sheet. This is to ensure that the asset value included in the balance sheet always denotes its lower bound. Thus, the balance sheet may understate the true value of the company's assets.[11]

   Asset reporting must address the issues of (a) asset ownership and (b) the certainty of its future economic benefits. If neither the ownership nor the future benefits are clearly established, an asset cannot be recognized. For example, companies routinely invest in employee training in the hope that doing so will lead to increased productivity at a future point in time. Since the completed training is really owned by the employees, and employees may leave at any time they wish, companies do not have real ownership of the training results. Thus, employee training is regarded as an expense and not an asset. When companies acquire plant facilities to make products for sales in the marketplace, its future benefits are more or less certain. Plant facilities are thus reported as assets. When companies apply resources to expand R&D and advertising, the future benefits of these investments are neither certain nor measurable. R&D and advertising are thus recognized as expenses and not assets.

   GAAP accounting rules in the United States contain one exception: Generally speaking, software development costs are to be reported as expenses as they are incurred. However, once the company management becomes confident that the software development efforts can be completed and the resulting software product will be used as intended, all costs incurred from that point on are to be reported as assets.

There are several ambiguities in the accounting rules practiced in the United States:

1. *Buying versus developing*: If Company A acquires Company B by paying a purchase price that exceeds Company B's net asset value, then this excess value is called goodwill. Goodwill includes the intangible assets of Company B such as its brand name, trademarks, patents, R&D portfolio, and employee skills. After the merger, the surviving company has part of its R&D (from the original Company A) recognized as expenses, and part of the R&D (acquired from Company B) as assets.
2. *Valuing intangible assets*: "If you can't kick a resource, it really isn't an asset." This saying is typically the justification used by companies to rapidly write off intangible assets from their balance sheets. Oftentimes, goodwill is significantly overvalued in a merger or acquisition transaction due to potential conflicts of interests among the parties involved. Writing off intangible assets distorts the true value of the assets reported in the company's balance sheet.
3. *Market value*: U.S. accounting rules prescribe that marketable securities (e.g., stocks, bonds, and real estates) are to be reported at their fair market values only if they are not to be held to maturity. Thus, at any given time, the real asset value of a company is distorted by not reporting the current true market values of these assets in balance sheets.

B. ***Liabilities*** are obligations to be satisfied by transferring assets or providing services to another entity (e.g., banks, suppliers, and customers).

A liability is recorded when an obligation has been incurred and the amount and timing of this obligation can be measured with a reasonable amount of certainty.[12] For multiple-year commitments, the recordable obligation is the present value of expected future commitments wherein the discount rate (interest rate) is the prevailing rate when the obligation was first established (see Appendix 2.10.2).

C. ***Revenue*** recognition must satisfy two conditions: Revenue is earned when (a) all or substantially all of the goods or services are delivered to the customers and (b) it is likely that the collection of cash or receivables will be successful. Generally speaking, the timing of product or service delivery may not be the same as that for payment collection.[13]

For magazine subscriptions, insurance policies, and service contracts, customers usually pay in advance. In these cases, payments received ahead of the service delivery dates are kept in a "deferred revenue" account. Only after the pertinent service is delivered will the applicable payments be credited to the revenue T-account during the accounting period. A detailed discussion on T-accounts is offered in Appendix 3.8.1. For products sold on credit, companies recognize the revenue as soon as the products are shipped out, and invoices are issued to customers ahead of the payment collection.

In the case of construction projects, which usually stretch out over many accounting periods, revenues are registered in T-accounts by using the *percentage completion* method and are recognized in proportion to the expenses incurred in the project.

For products sold with money-back guarantees, companies recognize revenue at the time the product is delivered. At the end of an accounting period, management makes an estimate of the cost of returns (a liability) to adjust the revenue figure.

D. ***Expenses*** are economic resources that either have been consumed or have declined in value during an accounting period. Expenses are typically recorded in the form of a reduction of asset value (e.g., cash) or by a creation of liability (e.g., accounts payable).

There are three types of expenses: (a) consumed resources having a cause-and-effect relationship with revenue generated during the same accounting period (e.g., CGS); (b) other resources consumed during the same accounting period, but having no cause-and-effect relationship with revenue (such as R&D expenses, advertising expenses, depreciation charges, local taxes, pension expenses, and other general administrative expenses); (c) reduction of expected benefits of company assets generated by past investments (e.g., the write-down of production facilities and equipment that is no longer of value due to the noncompetitiveness of products or technological obsolescence).[14]

3.3.6 CAUTION IN READING FINANCIAL STATEMENTS

In the United States, companies follow the GAAP defined by the industrial panel named Financial Accounting Standards Board (FASB), in preparing financial statements. Even so, all of these financial statements are not created equally; they need to be studied carefully because of the following built-in variations:

1. *Depreciation accounting base*: Some companies may use straight-line method, whereas others may choose to use the double-declining method, as both are allowable.
2. *Inventory accounting method*: Some companies use first in and first out (FIFO) methods, whereas others my choose to use last in and first out (LIFO). In time periods with high price volatility, the inventory value will be affected by the inventory accounting method chosen.
3. *Cost of capital*: Companies incur cost when employing capital, which is typically generated by selling company stocks (equity financing) and/or by issuing corporate bonds (debt financing). Dependent on the debt to equity ratio and cost of raising equity and debt, the weighted average of cost of captial (WACC) will be different from time to time and from one company to another.
4. *Difference between book and market values*: The book values of companies' fixed assets are calculated by substracting the accumulated depreciation from their initial acquisition prices. These values may be quite different from the assets' current market values, which in turn are dependent on the supply and demand in the marketpalce and the overall economy. Such differences may have to be empirically readjusted, when evaluating them.
5. *Long-term liabilities reportable*: The current FASB rules do not require companies to report long-term liabilities associated with pension, health care, and other such obligations in balance sheets, although these liabilities are typically disclosed in footnotes and other such obsure places within the companies' annual reports. Some companies with a heavily unionized workforce could therefore create the illusion of having a higher net worth than otherwise before uninformed investors. In March 2006, FASB annouced the intention of initiating new regulations to improve the transparency of this disclosure. It is thus important to keep such long-term liabilities in mind when reading financial statements of any company.

Example 3.3

Superior Technologies sells a product at the unit price of $100. The unit cost of the product is $60. Annual sales have averaged 1 million units, and its annual selling expense has been $7 million. Market research has determined that, if the selling price of the company product is decreased to $90, there will be a 35 percent increase in the number of units sold. The engineering department estimates that, if the production volume is increased by 35 percent, it will reduce the unit product cost by 10 percent due to the scale of economies. To pursue the 35 percent increase in sales volume, the company's selling expense will need to increase by about 50 percent.

The company's current warehouse facilities are sufficiently large to accommodate the possible increase of 35 percent in sales volume without requiring new investment. Furthermore, regardless of the product price, the company is obliged to pay an annual loan interest of $2 million. Its corporate tax rate is 45 percent. It maintains an R&D department, whose operation is independent of the sales units, at an annual cost of $5 million. Its administrative expense is $15 million, which is also independent of the sales activities. In addition, the company incurs a pretax depreciation charge of $2 million.

Determine if the reduction of product price would increase or decrease the net income of the company and by how much.

Answer 3.3

The reduction of product price will cause the company's net income to increase to $7.75 million from $4.95 million (see details in Table 3.7).

Table 3.7. Net income due to increased sales

Items	Current operation (dollars)	Operation with increased unit sales (dollars)
Units of product sold	1,000,000	1,350,000
Product price	100	90
Unit product cost	60	54
Sales revenue	100,000,000	121,500,000
Cost of goods sold	60,000,000	72,900,000
Gross margin	40,000,000	48,600,000
Selling expense	7,000,000	10,500,000
Administrative expense	15,000,000	15,000,000
R&D	5,000,000	5,000,000
Depreciation	2,000,000	2,000,000
EBIT	11,000,000	16,100,000
Interest	2,000,000	2,000,000
Taxable income	9,000,000	14,100,000
Tax (45%)	4,050,000	6,345,000
Net income	4,950,000	7,755,000

Example 3.4

The company is considering the introduction of a new product that is expected to reach sales of $10 million in its first full year of operation and $13 million of sales in the second and third years. Thereafter, annual sales are expected to decline to two-thirds of peak annual sales in the fourth year, and one-third of peak sales in the fifth year. No more sales are expected after the fifth year.

The CGS is about 60 percent of the sales revenues in each year. The sales general and administrative (SG&A) expenses are about 23.5 percent of the sales revenue. Tax on profits is to be paid at a 40 percent rate.

A capital investment of $0.5 million is needed to acquire production equipment. No salvage value is expected at the end of its five-year useful life. This investment is to be fully depreciated on a straight-line basis over 5 years.

In addition, working capital is needed to support the expected sales in an amount equal to 27 percent of the sales revenue. This working capital investment must be made at the beginning of each year to build up the needed inventory and to implement the planned sales program.

Furthermore, during the first year of sales activity, a one-time product introductory expense of $200,000 is incurred. Approximately $1.0 million had already been spent promoting and test marketing the new product.

A. Formulate a multiyear income statement to estimate the cash flows throughout its five-year life cycle.
B. Assuming a 20 percent discount rate, what is the new product's net present value (NPV)?
C. Should the company introduce the new product?

Answer 3.4

Table 3.8 presents the expected cash flows from year 1 to year 5. Its present value is $3.07 million.

Introducing this new product line requires a capital investment of $1.5 million and total working capital investment of $6.425 million over five years. Thus, the NPV of this new product line is estimated to be –$4.8 million. The new product line should not be introduced because of its projected poor profitability.

Table 3.8. Net present value analysis

Entries	Year 1 (thousands of dollars)	Year 2 (thousands of dollars)	Year 3 (thousands of dollars)	Year 4 (thousands of dollars)	Year 5 (thousands of dollars)
Interest rate = 0.2%					
Sales	10,000	13,000	13,000	8,667	4,333
CGS	6,000	7,800	7,800	5,200	2,600
Gross margin	4,000	5,200	5,200	3,467	1,733
SG&A	2,350	3,055	3,055	2,037	1,018
Depreciation	100	100	100	100	100
Expenses	200	0	0	0	0
EBIT	1,350	2,045	2,045	1,330	615
Tax (0.4%)	540	818	818	532	246
EBIAT	810	1,227	1,227	798	369
Cash flow	910	1,327	1,327	898	469
Present values	758.33	921.53	767.94	433.08	188.47
PV (cash flow) = 3,069.35					
Initial investment = 1,500					
WC investment	2,700	1,790	2,183	1,013.09	271.88
PV(WC) = 6,425.03					
NPV = - $4,855.68 (= 3,069.35 – 1,500 – 6,425.03)					

Notes: EBIAT = earnings before interest after tax; WC = working capital

3.4 FUNDAMENTALS OF FINANCIAL ANALYSIS

The purpose of conducting financial analyses is to assess the effectiveness of the company's management in achieving the objectives set forth by the company's board of directors with respect to a number of critically important business factors.[15] Such factors include:

1. **Liquidity**—the availability of current assets to satisfy the firm's operational requirements
2. **Activity**—the efficiency of resource utilization
3. **Profitability**—the extent of the firm's financial success
4. **Capitalization**—the makeup of the company's assets and its utilization of financial leverage
5. **Stock value**—the market price of the company's stock. The product of stock price and the number of outstanding stocks is defined as *market capitalization.*

Typically, the corporate objectives of many for-profit companies are growth, profitability, and return on investment (ROI).

The *growth* objective suggests that companies keep product prices low, increase marketing expenses, run plants at full capacity, take loans to keep inventory high, strive for a larger market share and gain a more dominant market position. The *profitability* objective dictates that companies set prices high to maximize profits, run plants at a capacity that minimizes costs (production and maintenance), and use debt, when called for, to reduce tax liabiities. The *ROI* objective is achieved by operating the company to maximize its financial return with respect to the firm's investment (e.g., the "milk-the-cash-cow" strategy, see Section 4.5.1[D]).

Financial analyses focus on studying period-to-period changes in key financial data (internal benchmarking) and on comparing performance ratios with the applicable industrial standards (external benchmarking).

3.4.1 PERFORMANCE RATIOS

In this section, we introduce a specific system of calculating performance ratios by grouping together the ratios for liquidity, activities, profitability, capitalization, and stock value.[16] In order to understand the performnce analysis system, we will want to first learn how each of these terms is defined.

A. **Liquidity:** It is the firm's capability to satisfy its current liabilities, such as buying materials, paying wages and salaries, and paying interests on long-term debt, and other necessary expenditures. Without liquidity, there can be no activity.

Working capital is defined as current assets minus current liabilities. The changes in working capital over several periods provide an indication of the company's reserve strength to weather financial adversities.

Current ratio is the ratio of current assets to current liabilities. Current assets are frequently considered the major reservoir of funds for meeting current obligations. This ratio provides an indication of the company's ability to finance its operations over the next 12 months. A current ratio above 1.0 indicates a margin of safety that allows for a possible shrinkage of value in current assets such as inventories and accounts receivables. However, having a current ratio in excess of 2.0 or 3.0 may indicate a poor cash management practice.

Quick ratio is the ratio of quick asset to current liabilities. *Quick asset* is defined as cash plus marketable securities and accounts receivable. This ratio is more severe than the current ratio in that it excludes the value of inventory whose liquid value may not be certain. Thus, the quick ratio indicates the company's ability to meet its financial obligations over the next 12 months without the use of inventory that may take time to unload. Sometimes, it is also called the *acid test* or *liquidity ratio*.

B. **Activity:** The changes in sales and inventory are activities. Successful activity leads to profitability.

Collection period ratio is the accounts receivable divided by average daily sales and it is measured in days. The average daily sales is the total annual sales divided by 360 days. This ratio measures the managerial effectiveness of the credit department in collecting receivables and the quality of accounts receivable.

Inventory turnover ratio is the CGS divided by the average inventory. It expresses the number of times during a year that the average inventory is recouped or turned over through the company's sales activities. The higher the turnover, the more efficient will be the company's inventory management performance, provided that there has been no shortage of inventory, which would result in a loss of sales and a failure to satisfy customers' needs.

Asset turnover ratio of net sales to total assets is a measure of the ability of the company's management to utilize total assets to generate sales.

Working capital turnover ratio is net sales to working capital. Working capital is defined as average current assets minus average current liabilities. It indicates the company's ability to efficiently utilize working capital to generate sales.

Sales to employee ratio is the company's net sales revenue divided by its average number of employees working during an accounting period. It measures the company's ability to effectually utilize human resources.

C. **Profitability:** To be profitable is the objective of all for-profit companies. Without liquidity and activity, there can be no profitability. If the company is profitable, it can readily obtain the required liquidity to keep its operations continuing.

Gross margin to sales ratio measures the company's profitability on the basis of sales. Gross margin is defined as sales revenue minus CGS. Gross margin percentage is defined as the gross margin divided by sales.

Net income to sales ratio indicates the company's overall operational efficiency (e.g., procurement, cost control, current assets deployment, and utilization of financial leverage) in creating profitability based on sales. This ratio is also known as ROS, which stands for return on sales.

Net income to OE ratio measures profitability from the shareholders' viewpoint. It is also known as ROE, which stands for return on equity. This ratio indicates the earning power of the ownership investment in the company.

Net income to total asset ratio is net income divided by total assets. It measures the management's ability to effectively apply company assets in generating profits. It is known as ROA, which stands for return on assets.

D. **Capitalization:** The sum of the company's long-term liabilities and OE is defined as the total capital deployed by company management to pursue business opportunities. Several ratios are in use to assess this capital deployment effectiveness.

Returned on invested capital is the ratio of net income divided by capital.

Interest coverage ratio (EBIT divided by the interest expense) calculates the number of times the company's EBIT covers the required interest payment for the long-term debt—an indication of the company's ability to remain solvent in the foreseeable future.

Long-term debt to capitalization ratio is the ratio of long-term debt to the firm's capital—the total permanent investment in a company, indicating the percentage of long-term debt in the company's capital structure, excluding current liabilities. It is a measure of the company's financial leverage. Keeping this ratio small (hence, large

OE percentage) may not always be the smart choice, as the company will forgo the use of low-cost debt with tax deductible interest payments to enhance profitability.

Debt to equity ratio is the ratio of total liability to OE. It also measures the company's financial independence and the relative stake of shareholders (insiders) and bondholders (outsiders). A low ratio indicates that the company is financially secure so far as the owners are concerned. A high ratio indicates that the firm may have difficulty borrowing money in the future.

E. **Stock value:** This is the market price or value of the company's stock as defined by the financial markets. The company's management is obliged to pursue proper business strategies in order to steadily raise their stock value.

EPS is the ratio of net income minus preferred stock dividends divided by the number of common stocks outstanding.

Price to earning ratio (P/E) is the ratio of the market price of common share to earning per share. This ratio is widely used by financial analysts to predict the future market price of a compnay's stock by estimating its future earning per share values in the upcoming periods. In 2013, many companies listed in Standard and Poor 500 index have their P/E ratios in the range of 12 to 20.

Market to book ratio is the ratio of market price of stock to the book value per share. More precisely, the total book value of a company is defined as the total assets minus intangible assets, minus total liability, and minus the equity of preferred stocks. The book value per share is then the total book value divided by the number of outstanding common shares.

Dividend payout ratio is the ratio of dividends per share divided by earnings per share. It indicates the percentage of annual earnings paid out as dividends to shareholders. The portion that is not paid out goes into the retained earnings account on the balance sheet.

Example 3.5

The XYZ Corporation has current liabilities of $130,000 with a current ratio of 2.5:1. Indicate whether the individual transactions specified below will increase or decrease the current ratio or the amount of working capital and by how much in each case. Treat each item separately.

1. Purchase is made of $10,000 worth of merchandise on account.
2. The company collects $5,000 in accounts receivable.

3. Repayment is planned of notes payable that are due in current period, with $15,000 cash from bank account.
4. The acquisition of a machine priced at $40,000 is paid for with $10,000 cash, and the lump-sum balance is due in 18 months.
5. The company conducts a sale of machinery for $10,000. Accumulated depreciation is $50,000, and its original cost is $80,000.
6. The company pays dividends of $10,000 in cash and $10,000 in stock.
7. Wages are paid to the extent of $15,000. Of this amount, $3,000 had been shown on the balance sheet as accrued (due).
8. The company borrows $30,000 for one year. Proceeds are used to increase the bank account by $10,000 to pay off accounts due to the supplier ($15,000) and to acquire the right to patents ($5,000).
9. The company writes down inventories by $7,000 and organization expenses by $5,000.
10. The company sells $25,000 worth (cost) of merchandise from stock to customers who pay in 30 days. Company has a gross margin of 40 percent.

Answer 3.5

The following formula is used to determine current assets and working capital:

Current ratio (CR) = current assets (CA) divided by current liabilities (CL)

CR = 2.50
CL = $130,000
CA = 2.5 × 130,000 = $325,000
WC = CA – CL = 325,000 – 130,000 = $195,000

1. Increase in inventory: $10,000
   Increase in accounts payable: $10,000
   CR = 335/140 = 2.39 (down by 0.11)
   WC = unchanged.
2. Increase in cash: $5,000
   Decrease in accounts receivable: $5,000
   CR = unchanged
   WC = unchanged
3. Decrease in cash: $15,000
   Decrease in notes payable: $15,000

CR = 310/115 = 2.70 (up by 0.2)
WC = unchanged

4. Increase in fixed assets: $40,000 (no effect)
   Decrease in cash: 10,000
   Increase in long-term notes: 30,000 (no effect)
   CR = 315/130 = 2.42 (down by 0.8)
   WC = $185,000 (down by $10,000)
5. Decrease in fixed assets: $80,000 (no effect)
   Decrease in accumulated depreciation: $50,000 (no effect)
   Decrease in surplus: 20,000 (no effect)
   Increase in cash: 10,000
   CR = 335/130 = 2.58 (up by 0.08)
   WC = $205,000 (up by $10,000)
6. Decrease in cash: $10,000
   Decrease in surplus: 20,000 (no effect)
   Increase in stock: 10,000 (no effect)
   CR = 315/130 = 2.42 (down by 0.8)
   WC = $185,000 (down by $10,000)
7. Increase in cash: $10,000
   Increase in note payable: 30,000
   Decrease in accounts payable: 15,000
   Increase in patents: 5,000 (no effect)
   CR = 335/145 = 2.31 (down by 0.19)
   WC = $190,000 (down by $5,000)
8. Decrease in inventories: $7,000
   Decrease in organizational expense: $5,000 (no effect)
   Decrease in surplus: 12,000 (no effect)
   CR = 318/130 = 2.45 (down by 0.5)
   WC = $188,000 (down by $7,000)
9. Decrease in profit and loss: $12,000 (no effect)
   Decrease in cash: 15,000
   Decrease in accruals: 3,000
   CR = 310/127 = 2.44 (down by 0.6)
   WC = $183,000 (down by $12,000)
10. Increase in accounts receivable: $41,667
    Decrease in inventory: 25,000
    CR = 341,667/130,000 = 2.62 (up by 0.12)
    WC = $211,667 (up by $16,667)

The working capitals are as follows: (1) unchanged, (2) unchanged, (3) unchanged, (4) $185,000, (5) $205,000, (6) $185,000, (7) $190,000, (8) $188,000, (9) $183,000, and (10) $211,667.

3.4.2 RATIO ANALYSIS

Ratios are useful tools of financial analysis.[17] Sometimes ratios of significant financial data are more meaningful than the raw data themselves. They also provide an instant picture of the financial condition, operation, and profitability of a company, provided that the trends and deviations reflected by the ratios are interpreted properly.

Ratio analyses are subject to two constraints: past performance and various accounting methods. *Past performance* is not a sure basis for projecting the company's condition into the future. *Various accounting methods* employed by different companies may result in different financial figures (e.g., inventory accounting, depreciation, etc.), rendering a comparison being not always meaningful between companies in the same industry.

Typically, accountants try to reconcile financial statements before conducting comparative analyses. Examples of adjustments that are frequently made include:

1. Adjusting LIFO inventories to a FIFO basis (Appendix 2.10.4).
2. Changing the write-off periods of intangible assets, such as goodwill, patents, trademarks, and so on (Appendix 2.10.3).
3. Adding potential contingency liabilities if lawsuits are pending (Section 2.6.5[B]).
4. Reevaluating assets to reflect current market values (Section 3.3.5[A]).
5. Changing debt obligations to reflect current market interest rates.
6. Restating reserves or charges for bad debts, warranties, and product returns.
7. Reclassifying operating leases as capital leases.

When performing ratio analyses, it is advisable to follow this set of generally recommended guidelines:

1. Focus on a limited number of significant ratios.
2. Collect data over a number of past periods to identify the prevalent trends.
3. Present results in graphic or tabular form according to standards (e.g.,industrial averages).
4. Concentrate on all major variations from the standards.
5. Investigate the causes of these variations by cross-checking with other ratios and raw financial data.

In financial literature, many of the ratios defined above are being systematically collected and published for U.S. companies in various industries. Sources of information on ratios and other financial measures are typically reported regularly and made available for use by the public, including the following:

- *Value Line® Investment Survey*
- Standard & Poor's Industrial Surveys
- Moody's Investors Services

Another such source is the Ohio University library, which has links to *Factiva, One Source*, and others: http://www.library.ohiou.edu/subjects/businessblog/financial-ratios/. Other commercial sources are accessible through the Internet. For the 500 stocks comprising Standard & Poor's index, five specific ratios—the current ratio, long-term debt to capital, net income to sales, ROA, and ROE—are published in a widely available special guide for 10-year, consecutive periods.[18]

Example 3.6

For the years 2011 to 2012, the financial statements of XYZ Company are given in Tables 3.9 and 3.10. Define the performance ratios and compare them with industrial standards.

Answer 3.6

The 2011 performance ratios of XYZ Corporation are displayed here:

1. **Liquidity**
   (a) Current ratio = CA/CL = 3.1
   From the creditor's standpoint, this ratio should be as high as possible. On the other hand, prudent management will want to avoid the excessive buildup of idling cash or inventories (or both).
   (b) Quick ratio = quick asset/CL = 0.931
   A result far below 1:1 can be a warning sign.
   (c) Interest coverage ratio = 7.833
   The EBIT of the firm could pay 7.833 times the interest and other costs associated with the long-term debts. This ratio is good.

Table 3.9. XYZ balance sheet

Entries	2012 (thousands of dollars)	2011 (thousands of dollars)
Assets		
Cash	18,500	17,000
Marketable securities	0	5,000
Accounts receivable	39,500	28,500
Inventories	98,000	113,000
Total current assets	156,000	163,500
Plant and equipment (net)	275,000	290,000
Other assets	3,000	8,000
Total assets	434,000	461,500
Liabilities		
Accounts payable	34,500	18,000
Notes payable	20,000	25,000
Accrued expenses	18,500	11,500
Total current liabilities	73,000	54,500
Mortgage payable	20,000	30,000
Equities		
Common stock	200,000	200,000
Earned surplus	141,000	177,000
Total liabilities and equities	434,000	461,500

2. **Debt versus equity**
   - (d) Long-term debt to capitalization ratio = 4.44 percent. This debt level is prudent for firms in this industry.
   - (e) Total debt to owners' equity = 22.4 percent
     Total debt = CL + long-term debt
     OE = common stock + capital surplus + accumulated retained earnings
   - (f) Total debt to total asset ratio = 18.3 percent
3. **Activity**
   - (g) Sales to asset ratio = 0.86
   - (h) Ending inventory to sales ratio = 28.6 percent
   - (i) CGS/average inventory = 2.65 times

Table 3.10. XYZ income statement

Entries	**2012 (thousands of dollars)**	**2011 (thousands of dollars)**
Sales	330,000	395,000
Cost of sales*	265,000	280,000
Gross profit	65,000	115,000
Selling and administrative	95,000	88,000
Other expenses	4,000	3,500
Interest	2,000	3,000
Profit before taxes	–36,000	20,500
Federal income tax	0	10,000
Net income (loss)**	–36,000	10,500

*Includes depreciation of $15,500 in 2011 and $15,000 in 2012.
**No dividends paid in 2012.

Average inventory = the average of ending inventory of two consecutive years (e.g., 2001 and 2002)

4. **Profitability**
   - (j) Net income to owner's equity ratio = 2.8 percent
   - (k) Net income to sales ratio = 2.66 percent
   - (l) Gross margin to sales ratio = 29.1 percent
   - (m) EBIT to total asset ratio = 5.1 percent
   - (n) Net income to total asset ratio = 2.3 percent
   - (o) EBIT to sales ratio = 5.9 percent

Selected investment services companies (e.g., Value Line, Standard & Poors, Moody's) track various financial measures and report them regularly for use by the public. Engineering managers should develop a habit of reading and considering such reports.[19]

3.4.3 ECONOMIC VALUE ADDED

Developed by Stern Stewart & Company in 1989, economic value added (EVA) is an improved valuation method for asset-intensive companies or projects. EVA is defined as the after-tax-adjusted net operating income of a company or unit minus the total cost of capital spent during the same accounting period. It is also equal to the return on capital (ROC) minus the cost of capital, or the economic value above and beyond the cost of capital.[20] Sometimes EVA is also called "economic profit." In equation form, it is defined as:

$$\text{EVA} = \text{NOPAT} - \text{WACC} \times (\text{capital deployed}) \tag{3.2}$$

where

NOPAT = net operating profit after tax (net income)
Capital deployed = total assets – current liabiliites
WACC = weighted average of cost of capital (equity and debt) employed in producing the earnings (Section 2.6.3).

If EVA is positive, the company or unit is said to have added positive shareholder value. If EVA is negative, the company or unit is said to have diminished shareholder value.

EVA may also be applied to a single project by calculating the after-tax cash flows generated by the project minus the cost of capital spent for the project. For example, the NPV equation can be modified as follows:

$$\text{NPV} = -P + \left[\sum_{t=1}^{N} \frac{C_t - P \times \text{WACC}}{(1+\text{WACC})^t} \right] + \frac{\text{SV}}{(1+\text{WACC})^N} \tag{3.3}$$

Here

P = investment capital (dollars) for the project
C_t = net after-tax cash inflow (dollars) to be produced by the project in period (t)
$P \times$ WACC = cost of capital (dollars) spent during the period (t)
WACC = weighted average cost of capital (percent) in effect
SV = salvage value of the project at the end of N periods
N = number of period

The major advantage of EVA over ROC is that it may encourage managers to undertake desirable investments and activities that will increase the value of the firm. The difference between these two methods is illustrated by the ABC Company example shown below:

ABC company has established that its WACC is 10 percent, and its ROC standard for investment purposes is 14 percent. The management is considering a new capital investment that is expected to earn a return of 12 percent. This new investment is attractive according to the EVA criterion, as 12 percent is larger than 10 percent. However, this new investment is a poor choice if evaluated with the ROC criterion, because 12 percent is less than 14 percent. Thus, by using the ROC criterion to evaluate investments, the company may lose the opportunity to create shareholder value.

A 10-year study has shown that there is a general correlation between EVA and stock returns of innumerable companies. However, the correlation between EVA and wealth creation (in the form of stock price increase) is weak. Among users of EVA, the following firms are known leaders in industry*: AT&T, Eastman Chemical, Coca-Cola, Eli Lilly, and Wal-Mart.*

Engineering managers should learn to apply EVA in order to strengthen their financial accounting skills.

Example 3.7

Define the EVA of XYZ Company (see its income statement in Table 3.1 and its baalance sheet in Table 3.4), by assuming that the company's WACC is 12.35 percent for both years. Discuss the EVA results.

Answer 3.7

For the year 2011, the NOPAT for XYZ Company is \$559.60. Its invested capital is total assets minus current liabilities (\$8,155 – \$1,446.90) and has the value of \$6,708.10. Thus, EVA = 559.60 – 0.1235 × 6,708.10 = –\$268.85 million. For the year 2012, the NOPAT for XYZ is 410.3, and its invested capital is 7,530.5 (=\$9,167.90 – 1,637.4). Its EVA = 410.3 – 0.1235 × 7530.5 = –\$519.72 million.

The negative values for both years were not obvious without EVA accounting. By performing "what-if" analyses of XYZ's income statement, company management can readily determine what it will take to improve its EVA values.

3.4.4 CREATION OF SHAREHOLDER VALUE

Shareholders are primarily interested in the total return to stockholders (TRS), which consists of the yield of dividend paid out by the company plus the rate of the stock's long-term (e.g., 10 years) appreciation potential.[21] The amount of annual dividend paid out is a function of the net income earned by the company in a given year. Many established companies strive to continously pay dividends in order to appease investors. Being able to pay dividend quarter after quarter regularly requires that the company focuses on its short-term operations.

The long-term appreciation rate of the company's stock, on the other hand, depends on the company's long-term investment strategies. The stock price is an indication of the market's assessment of the company's future expected cash flows. These future expected cash flows are to be created by projects that add value to the company over a long period of time. The finance function is particularly qualified to offer advice in (a) increasing spending in value-creating activities (such as technology development, product/service design, marketing, advertising, etc.), (b) systematically reviewing and retaining only assets that maximize value creation, and (c) pursuing proper acquisition and divestiture strategies in order to maximize the creation of this value.

Example 3.8

The Global Business Services (GBS) organization within Procter & Gamble (P&G) provides services in finance, accounting, employee services, customer logistics, purchasing, and IT to other divisions of the company. Over the years, P&G has prided itself on its inclusive company culture and progressive employment practices (e.g., job guarantees, length of employment, and salary/benefit policies favorable to employees). In an attempt to maximize shareholder value, P&G management periodically reviews its resource deploying strategies. Since GBS is not one of the company's core competencies, there has not been enough management attention, and consequently resources devoted to its continuous improvement.[22] Eventually, questions have been raised regarding the service quality and costs of the GBS organization to P&G. Finally, a management committee has recommended that P&G consider one of the following options:

1. Spin off GBS to be an independent company that continues to provide services to P&G as a subcontractor.
2. Outsource all services to one outside company and transfer all employees over.

3. Outsource services to a number of separate providers in order to enhance quality.
4. Continue GBS in-house.

Which one is the right option for P&G to take in this case?

Answer 3.8

The best choice is to make use of the Kepner-Tregoe method[23] by (a) first specifying a number of mutually exclusive and collectively exhaustive decision criteria, (b) assigning weight to each criterion to indicate its relative importance, (c) ranking all options against each of these criteria, and finally (d) computing the weighted scores, which indicate a relative ranking of the options under consideration.

An answer to Example 3.8 may be obtained by selecting a set of decision criteria (five in this case) as shown in column 1 of Table 3.11. The relative importance of these criteria are specified by the choices of the weight factors (see column 2 of Table 3.11). All options are then evaluated against a specifc criterion. For criterion 1, the results of this evaluation are shown in row 3 of Table 3.11. The scores 10, 8, 6, and 3 are assigned to indicate that for criterion 1, options are respectively ranked in the order of 4, 2, 3, and 1, from most satisfactory to least satisfactory. This process of evaluation continues for the remaining criteria. Afterward, the total weighted score

Table 3.11. Application of the Kepner-Tregoe method

Criteria	Weight	Option 1	Option 2	Option 3	Option 4
1. Compatibility with P&G policy and tradition	8	3	8	6	10
2. Service cost	9	10	6	8	3
3. Service quality	10	6	8	10	3
4. Value to affected P&G employees	8	8	6	3	10
5. P&G control	8	8	6	3	10
Total weighted score		302	294	268	297

is computed for each option. For option 1, the total weighted score is 302 (= 3 × 8 + 10 × 9 + 6 × 10 + 8 × 8 + 8 × 8). Similarly, the total weighted score for the other options can be readily calculated to be 294, 268, and 297, respectively, as shown in the last row of Table 3.11.

An examination of the total weight scores indicates that option 1 should be the preferred choice. An important advantage of applying the Kepner-Tregoe method in making decisions is that both the decision criteria and evaluation score are transparently defined to enhance their understanding and acceptability by others who might be affected by the decision results.

Example 3.9

The 2000 to 2001 income statement and balance sheet for Buffalo Best are presented in Tables 3.12 and 3.13. The WACC for Buffalo Best is 10 percent for both years.

Table 3.12. Income statement of Buffalo Best

Entries	Year 2012 (thousands of dollars)	Year 2011 (thousands of dollars)
Sales	18,000	17,000
Cost of goods sold	11,000	10,500
Gross margin	7,000	6,500
Administrative and selling expenses	3,500	3,200
R&D	500	500
Depreciation	1,000	1,000
EBIT	2,000	1,800
Interest	100	100
Taxable income	1,900	1,700
Tax (30%)	570	510
NOPAT	1,330	1,190
Dividends	330	190
Retained earnings	1,000	1,000

Table 3.13. Balance sheet of Buffalo Best

Entries	Year 2012 (thousands of dollars)	Year 2011 (thousands of dollars)
Assets		
Cash and securities	5,000	6,000
Accounts receivable	15,000	10,000
Inventory	10,000	7,300
Net fixed assets (investment minus accumulated depreciation)	50,000	51,000
Other	1,000	1,200
Total assets	81,000	75,500
Liabilities		
Accounts payable	20,000	15,000
Other short-term liabilities	26,000	24,000
Long-term liabilities	1,000	1,500
Total liabilities	47,000	40,500
Equities		
Equities at par value	1,000	1,000
Capital surplus	12,000	14,000
Retained earnings	21,000	20,000
Total liabilities and owners' equities	81,000	75,500

Review and comment on the performance of the company based on the following:

A. Liquidity, activity, and profitability
B. Uses and sources of funds
C. Value creation based on EVA analysis

Answer 3.9

Tables 3.14 and 3.15 contain the answers to these questions.

Table 3.14. Performance ratios

Entries	2012	2011	Comments
1. Liquidity			
Current ratio	0.674	0.628	Poor liquidity, with current ratio less than 2.0 and negative working capital
Total liability to owner's equity	1.382	1.157	This number is acceptable being slightly higher than 1.0
Total debt to asset ratio	0.58	0.536	Total debt level could be increased somewhat.
Long-term debt to capitalization	0.029	0.041	Low-cost long-term debt is low, indicating a poor financing strategy.
Interest coverage ratio	20	12	This ratio is adequate.
2. Activity			
Collection period (days)	300	212	Poor credit policy and accounts management practices
Cost of goods sold to inventory	1.27	1.21	The inventory turns are small, keeping too high an inventory
Sales to asset ratio	0.222	0.225	Asset utilization to generate sales is reasonable.
Inventory to sales ratio	0.556	0.429	Too high inventory for the sales level achieved.
3. Profitability			
Gross margin to sales ratio	0.389	0.382	This ratio is reasonably good, indicating adequate production cost management
NOPAT to sales ratio	0.074	0.07	Small ratio indicates a need for better control of company expenses
NOPAT to owner's equity	0.039	0.034	Return for owners is poor, needing improvement.
NOPAT to total assets	0.0164	0.0158	Utilization of company assets is ineffective.
EBIT to sales ratio	0.111	0.106	Low ratio indicates a need for better controlling of company's general expenses

Table 3.15. Funds flow and EVA analyses

Sources of funds	**Dollars**	**Percentage**
Decrease in Cash & Securities	$1,000,000	9.50%
Depreciation	$1,000,000	9.50%
Decrease in Other Assets	200,000	1.90%
Increase in Accounts Payables	5,000,000	47.48%
Increase in other short-term liabilities	2,000,000	18.99%
Net Income	1,330,000	12.63%
Total	$10,530,000	100%
Use of Funds		
Increase in Accounts Receivables	5,000,000	47.48%
Increase in Inventory	2,700,000	25.64%
Decrease in Long-term Liabilities	500,000	4.75%
Increase in Retained Earnings	1,000,000	9.50%
Paying Dividends	330,000	3.13%
Increase in Retained Earnings	1,000,000	9.50%
Total	10,530,000	100%
EVA Analysis		
EVA(2001) =	–2,170,000	
EVA(2000) =	–2,460,000	

The sources' total is $10.53 million. The EVA was –$2.17 million in 2012 and –$2.46 million in 2011.

3.5 BALANCED SCORECARD

Financial ratios are developed to illustrate the companies' financial performance. Nonfinancial ratios are limited in number and restricted in scope, such as accounts receivable, collection period, inventory, utilization of fixed assets, and working capital.

All financial ratios are determined by using past performance data; they are "trailing" indicators, and as such they cannot foretell the future performance of a company. Because financial ratios are oriented to the short term, usually from one accounting period to another, company management is inadvertently forced to overemphasize short-term financial results, oftentimes neglecting the strategies for pursuing the company's long-term growth. The narrow focus of these financial ratios makes them no longer completely relevant to today's business environment, in which deployment of new technologies, customer satisfaction, employee innovation, and continuous betterment of business processes are key elements of company competitiveness in the marketplace.

These basic deficiencies in financial ratios are well recognized in industry. Attempts have been made in the past to modify these ratios as corporate measurement metrics. Kaplan and Norton[24] suggest a new set of corporate measurement metrics, the balanced scorecard, to cover four areas: (a) financial—shareholder value; (b) customers—time, quality, performance and service, and cost; (c) internal business processes—core competencies and responsiveness to customer needs; and (d) innovation and corporate learning—value added to the customer, new products, and continuous refinement. The significance of the suggested balanced scorecard lies in its balanced focus on both short-term profitability as well as long-term corporate growth. As the saying goes: "What you measure is what you get." Kaplan and Norton[25] advocate the use of a total of 15 to 20 metrics to cover these four areas to guide the company as it moves forward.

As an illustrative example, the balanced scorecard metrics for a typical manufacturing company may contain the following:

A. **Financial**—cash flow, quarterly sales growth and operational income, increased market shares, and ROE
B. **Customer**—percentage of sales from new products, percentage of sales from proprietary products, on-time delivery as defined by customers, share of key account's purchase, ranking by key accounts, and number of collaborative engineering efforts with customers
C. **Internal business process**—manufacturing capabilities versus competition, manufacturing excellence (cycle time, unit cost, and yield), design engineering efficiency, and new product introduction schedule versus plan
D. **Innovation and learning**—time to develop next-generation technology, speed to learn new manufacturing processes, percentage of products that equal 80 percent of sales, and new product introduction versus competition.

In general, balanced scorecard metrics for a given company must be built up according to its corporate strategy and vision, using a top-down approach. Doing so will ensure that performance metrics at lower management levels are properly aligned with the overall corporate goals. A unique strength of balanced scorecard metrics is that they link the company's long-term strategy with its short-term actions. These metrics contain forward-looking elements at the same time that they balance the internal and external measures. The creation of such metrics provides clarification, consensus, and focus on the desired corporate outcome.

According to Kaplan and Norton,[26] *United Parcel Service* has achieved an increase of 30 to 40 percent in profitability with balanced scorecard metrics. *Mobil Oil's North American Marketing and Refining Division* raised its standing from last to first in its industry after having implemented a balanced scorecard. Catucci[27] recommends that managers, when implementing a balanced scorecard, take personal ownership, nurture a core group of champions, educate team members, keep the program simple, be ruthless about implementation, integrate the scorecard into their own leadership systems, orchestrate the dynamics of scorecard meetings, communicate the scorecard widely, resist the urge for perfection, and look beyond the numbers to achieve cultural transformation of the company.

A widespread use of broad-based metrics, such as those suggested by the balanced scorecard, is likely to shift the attention of corporate management from a focus primarily on short-term financial performance to other areas of equal importance, such as customers, internal business processes, and innovation and learning. Contributions by engineering managers and professionals made in these nonfinancial areas will likely become readily and more favorably recognized in the future.

3.6 CAPITAL ASSETS VALUATION

Financial management deals with three general types of capital assets valuation problems: (a) assets in place (operations), (b) opportunities (R&D and marketing), and (c) acquisitions or joint ventures.

Capital budgeting problems related to assets in place are those that deliver a predictable string of cash flows in the immediate future. Examples include building a new plant facility, developing new products, and entering a new regional market. Sometimes these problems are grouped under the heading of operations, as investment in operations usually leads to immediate cash flows. Problems related to opportunities arise from

Table 3.16. Recommended capital asset valuation methods

Valuation problems	Recommended methods	Alternative methods
1. Assets in place (operations)	Adjusted present value (APV)	Multiples of sales, cash flows, EBIT, or book value; DCF (based on WACC), Monte Carlo simulations
2. Opportunities (R&D, marketing)	Simple option theory	Decision tree, complex option pricing, simulations
3. Equity claims	Equity cash flow	Multiples of net income; P/E ratios; DCF (based on WACC minus debt), simulations

decisions that do not generate an immediate flow of cash, but preserve a likelihood that future gains may be realized. Examples include R&D and marketing efforts. The third type of problems is related to acquisition, joint venture, formation of supply chains, and others, all of which may require the company to participate in equity investment and to share future equity cash flows with its business partners.

Generally speaking, each of these types of valuation problem is best handled by different valuation methods. Luehrman[28] suggests a list of recommended methods, as shown in Table 3.16.

3.6.1 OPERATIONS—ASSETS IN PLACE

There are several evaluation methods currently in use to assess capital projects in the investment category of operations.

A. Discount Cash Flow (Based on WACC)

Since 1980 or so, most companies have been using the DCF method to determine the NPV of an operation with assets in place and WACC to specify the discount factor: WACC stands for weighted average cost of capital, which represents average cost of money raised by the company through equity (selling stocks) or debts (borrowing from creditors such as banks and bond holders). Dependent on the specific way the company elects to acquire capital, this number (say 15 percent) is defined annually by the company's finance department.

$$\mathrm{NPV} = -P + \sum_{t=1}^{N} \frac{C_t}{(1+\mathrm{WACC})^t} + \frac{\mathrm{SV}}{(1+\mathrm{WACC})^N} \quad (3.4)$$

Here

NPV = net present value (dollars)
P = initial capital investment (dollars)
C_t = cash flows = future net benefits (dollars)
SV = salvage value = capital recovery (dollars)
N = total number of periods (year)
WACC = weighted average cost of capital (percent)

The NPV is equal to the present value of all future net benefits (e.g., income minus relevant costs), plus discounted capital recovery, if any, and minus the initial investment capital. It represents the net financial value added to a firm by a given capital investment project. Projects with large positive NPV values are favored. This method is sometimes called the *DCF (based on WACC) analysis*, as it is based on the use of the weighted cost of capital as the all-important discount factor.

Companies accept capital projects if the NPV is greater than zero, meaning that an initiation of such projects will lead to net positive value being added to the companies.

B. Internal Rate of Return

A popular variation to DCF (based on WACC) is the internal rate of return (IRR). When applying Equation 3.4 to evaluate projects, IRR is the discount rate that is realizable when the present values of all DCFs balance the initial investment (by setting NPV equal to zero). IRR represents the reinvestment rate, which is held constant over the duration of a project. For example, assuming no salvage value, the IRR of a $10,000 investment that yields a revenue of $5,000 per year for three years is 23.35 percent. Even though the concept is logical and its computation remains straightforward, IRR does not always indicate the true annual ROI.[29] This is especially the case when the interim cash flows produced by the invetmant at hand can only be reinvested at a rate lower than IRR. For most projects, these interim cash flows may be realistically reinvested at WACC, the company's weighted averaged cost of capital. If WACC is lower than IRR, then the computed IRR overestimates the true ROI. Thus, the use of IRR could lead to wrong investment decisions and budget distortions, if the reinvestment rate of the interim cash flow is lower than the computed IRR.

In general, companies specify a hurdle rate that must be met or exceeded by the IRRs of all acceptable capital projects. By adjusting the hurdle rate according to conditions related to market, economy, and environment for a specific period involved, companies exercise control over the capital investment standards. The hurdle rate is typically three to four times of WACC in value.

C. Multipliers

Another method to estimate the proper capital investment in a project is to use numerical multipliers that are based on historical data. Specifically, average multipliers are defined for use in conjunction with commonly available financial data such as sales, book value, EBIT, and cash flow.

In general, the financial data of many publicly held U.S. companies in various industries are widely available in the literature, including publications such as Standard & Poor's industry surveys, and Value Line Industrial surveys. Sales figures of numerous companies are readily obtained, and their relationship with company assets is typically recorded as the *asset turn ratios* (see Section 3.4.1). The reciprocal of this ratio is a multiplier that, when used together with the known sales figure, provides a rough estimate of the company's asset value.

For the XYZ Company described in Tables 3.1 and 3.4, the sales to total asset ratio for the year 2011 is 1.0276 (=8,380.30/8,1555.0). The reciprocal of this number is 0.9731, which is the sales multiplier to determine assets. If a sufficient number of other companies in the same industry are surveyed, the resulting average industry-based multiplier can be used to generate a preliminary estimate of the assets employed to produce a known sales revenue.

To determine the capital investment value of a plant expansion, new product development, and other products, the company's existing total sales to total asset ratio may serve as a good yardstick to ascertain a reasonable capital investment level, but only if the project outcome in terms of future sales can be estimated.

How much debt should the company incur to finance a specific project? The *debt to asset ratio* is linked to the debt to equity ratio (see Section 3.4.1). For the XYZ Company described in Tables 3.1 and 3.4, the debt to asset ratio for 2011 is 0.405 (=3,304.80/8,155.00). Again, if an industrial average is found for this multiplier, it can serve as a useful tool to estimate a reasonable debt level for a project on the basis of its known asset value.

Similarly, industrial average multipliers may be found for cash flow. Cash flow for a given accounting period is defined as

$$\text{Cash} = \text{NI} + \text{Dep} \tag{3.5}$$

where NI is net income and Dep is depreciation charge invoked in a given accounting period; both are entries in a typical income statement.

Note that financial charges (e.g., interest payment) are usually ignored in the cash flow computation. Depreciation, a noncash expense, is added back. The following numerical example offers additional clarification; see Table 3.17.

$$\begin{aligned} \text{NI} &= 200{,}000 - 80{,}000 - 40{,}000 - 10{,}000 - 5{,}000 - 12{,}000 = 53{,}000 \\ \text{Dep} &= 20{,}000 \\ \text{Cash} &= \text{NI} + \text{Dep} = 53{,}000 + 20{,}000 = 73{,}000 \end{aligned}$$

For the XYZ Company in 2011, the cash flow is $856.8 million (=559.6 + 207.2), calculated from data contained in Table 3.1. Yet another multiplier is related to EBIT. This multiplier is estimated to be 8.996, being the reciprocal of the company's EBIT to asset ratio of 0.111 (=906.5/8,155.0) as evident in Tables 3.1 and 3.4.

Table 3.17. Sample data set

Sales	$200,000
Manufacturing costs (including a depreciation of $20,000)	80,000
Sales and administrative expenses	40,000
Equipment service charges	10,000
Decrease in contribution of existing product	5,000
Increase in accounts receivable	15,000
Increase in inventory	20,000
Increase in current liability	30,000
Income tax	12,000
Interest paid for bonds used to finance projects	18,000

The use of some of these multipliers in combination is likely to generate figures that can serve as acceptable ballpark estimates of the required investment for a new project.

D. Monte Carlo Simulations

Monte Carlo simulations refer to a sampling technique that processes input data presented in the form of distribution functions. All input variables are simultaneously varied within each of their respective ranges, as defined by their distribution functions. The mathematical operations (e.g., addition, subtraction, multiplication, and division) of a given cost model are first defined in spreadsheet programs[30] (e.g., Excel). Upon activating a suitable risk-analysis software program, Monte Carlo simulations produce one or more outputs that are also in the form of distribution functions[31] (see Section 2.6.3).

In evaluating operations, Monte Carlo simulations may be usefully applied to the DCF (based on WACC) method. All future net cash flows and discount rates are modeled by distribution functions (e.g., triangular or Gaussion), as they are indeed expected to vary within ranges. The DCF (based on WACC) equation is readily set up in a spreadsheet program. As the sole output, the NPV will also vary within a lower and upper bound. The following results will be helpful to decision makers of capital budgeting:

1. The maximum probability at which the NPV is projected to be negative
2. The probability at which the NPV is projected to exceed a given value (e.g., $10 million)
3. The standard deviation of the NPV output
4. The minimum NPV
5. The maximum NPV

Monte Carlo simulations are also applicable to the calculation of IRR for the evaluation of operations.

3.6.2 OPPORTUNITIES—REAL OPTIONS

The second category of valuation problems is related to opportunities such as R&D and marketing projects. These problems do not lend themselves to DCF analyses, which require the estimates of projected future cash flows. If there is no cash flow, there can be no net positive present value. Financial analysts and researchers in the literature recommend that the *European simple call option* method be used to price these opportunities.

Option is a common tool frequently used for trading securities in the financial markets. There are *call* and *put* options. A *call option* provides its holder with the right, but not the obligation, to buy 100 shares of an underlying company (e.g., Apple or IBM) by a certain expiration date (typically three months from the present) at a specific price (*strike* or *exercise price*). The holder pays a fee, or premium, to buy the call option, which he or she may exercise on any business day up to and including the contract expiration date.

A *put option* gives its holder the right, but again not the obligation, to sell 100 shares of a given stock within a period of time at a predetermined strike price. Investors who predict that the stock price of a given company is going to decline in the future will want to sell the stocks today and buy a call option to recover the stocks at a lower price in the future. The premium for an option is dependent on five factors: (a) current stock price, (b) strike price, (c) length of option contract, (d) stock price volatility, and (e) current opportunity cost (e.g., bank interest).

For the purpose of evaluating capital project opportunities, the European simple call option is more appropriate in that the call option can be exercised only on the expiration date specified in the contract and no sooner. Table 3.18 shows the equivalence between financial calls and real calls.

Companies with new technologies, product development ideas, defensive positions in fast growing markets, and access to new markets have valuable opportunities to explore. When dealing with opportunities,

Table 3.18. Equivalence of call options

Financial calls	Real calls
Current stock price	Underlying asset value
Length of contract	Length of time to invest
Volatility	Volatility of future project value
Strike price	Capital investment for the project
Current bank interest	Cost of equity capital

three possible scenarios exist: (a) to invest immediately, (b) to reject the opportunity right away, and (c) to preserve the option of investing in the opportunity at a later time.[32]

This problem may be studied by using the Black-Scholes option-pricing model (BSOPM).[33] The BSOPM is defined as

$$C = V\left[N(d)\right] - e^{-rT} \times \left[N\left(d - \sigma\sqrt{T}\right)\right] \quad (3.6)$$

$$d = \frac{\ln\left(\frac{V}{X}\right) + T\left(r + \frac{\sigma^2}{2}\right)}{\sigma\sqrt{T}} \quad (3.7)$$

where

C = option price
V = current value of the underlying asset
X = the exercise price of the option
σ = annual standard deviation of the returns on the underlying asset
r = the annual risk-free rate
$N(d)$ = cumulative standard normal distribution function evaluated at d
$\ln(x)$ = natural log function of x
T = time to expiration of the option (years)

Table 3.19 exhibits the representative data of the $N(d)$ function.

Example 3.10

If $1 million is invested immediately, there will be a loss of $100,000 due to the current economic condition and marketing environment. However, if the company waits for two years, things may be different. What should the company do, assuming that the current risk-free rate is 7 percent and the project volatility is 0.3?

Answer 3.10

The problem may be studied by using BSOPM. Let us define the following equivalents:

Table 3.19. Values of cumulative normal distribution function

d	*N*(*d*)	*d*	*N*(*d*)	*d*	*N*(*d*)
–3.00	0.0013	–1.00	0.1587	1.00	0.8413
–2.90	0.0019	–0.90	0.1841	1.10	0.8643
–2.80	0.0026	–0.80	0.2119	1.20	0.8849
–2.70	0.0035	–0.70	0.242	1.30	0.9032
–2.60	0.0047	–0.60	0.2743	1.40	0.9192
–2.50	0.0062	–0.50	0.3085	1.50	0.9332
–2.40	0.0082	–0.40	0.3446	1.60	0.9452
–2.30	0.0107	–0.30	0.3821	1.70	0.9554
–2.20	0.0139	–0.20	0.4207	1.80	0.9641
–2.10	0.0179	–0.10	0.4602	1.90	0.9726
–2.00	0.0228	0.00	0.5	2.00	0.9772
–1.90	0.0287	0.10	0.5398	2.10	0.9821
–1.80	0.0359	0.20	0.5793	2.20	0.9861
–1.70	0.0446	0.30	0.6179	2.30	0.9893
–1.60	0.0548	0.40	0.6554	2.40	0.9918
–1.50	0.0668	0.50	0.6915	2.50	0.9938
–1.40	0.0808	0.60	0.7257	2.60	0.9953
–1.30	0.0968	0.70	0.758	2.70	0.9965
–1.20	0.1151	0.80	0.7881	2.80	0.9974
–1.10	0.1357	0.90	0.8159	2.90	0.9981

Given are the following data:

$$\sigma = 0.30$$
$$r = 0.07$$
$$X = \$1{,}000{,}000$$
$$V = \$900{,}000$$
$$T = 2$$

From Equation (3.7),

$$d_1 = \frac{\left\{ \ln(900{,}000/1{,}000{,}000) + 2\left[\left(0.07 + \frac{0.3^2}{2} \right) \right] \right\}}{0.3\sqrt{2}} = 0.2938$$

$$d_2 = 0.2938 - 0.3 \times 1.41456 = -0.1306$$

From Table 3.19, we get by interpolation:

$$N(0.2938) = 0.6155$$
$$N(-0.1356) = 0.4462$$

From Equation 3.6, the option price becomes

$$C = 900{,}000 \times 0.6155 - \exp(-0.07 \times 2) \times 1{,}000{,}000 \times 0.4462$$
$$= \$166{,}042$$

Now the company has the alternative of investing $166,042 to preserve investment opportunities for two years, in addition to deciding for or against it right away. If this option is preserved, additional information that will cause a change in the underlying asset value may become available during the ensuing two years. The company may still decide not to invest in two years, but preserving the option to invest is a valuable alternative.

3.6.3 ACQUISITION AND JOINT VENTURES

Companies do constantly assess the needs of applying new technologies to enhance the competitiveness of their products/servcies. Instead of relying on their in-house efforts, acquisition of other companies having unique technical strength is a frequently exercised option.

When considering specific companies as candidates for possible acquisition, the evaluation of the assets of the target companies becomes critically important. A number of methods are practiced to assess the value of a target company.[34]

The value of a company is defined by its equity and debt, that is,

$$V = E + D \qquad (3.8)$$

where

V = firm's value in the market

E = equity (stocks)

D = debt outstanding (i.e., bonds, loans, etc.)

The company management continues to maximize the target firm's value for its shareholders. Two factors may affect this value maximization attempt: (a) *takeover bids* (when acquiring firm is enticed to pay a higher than normal premium to absorb the acquisition candidates) tend to raise the stock price; and (2) *stock options* (rights to buy stock at a fixed price awarded to company's management personnel and new hires) tend to dilute the shareholder value.

Example 3.11 illustrates one possible ways of evaluating the net value of a target company, when defining the bid price for such an acquisition.

Example 3.11

XYZ Company is considering the acquisition of Target Company, a smaller competitor in the same industry. The income statement of Target Company is displayed in Table 3.20. As a stand-alone company, its sales, CGS, depreciation, selling, and administrative expenses are all projected to increase by 4 percent per year.

To maintain the projected sales growth of Target Company, XYZ Company must make additional working capital investments (see Table 3.20).

A. Assuming a 10 percent discount rate, what is the maximum price XYZ Company should be willing to pay for this acquisition if it is to be run as a stand-alone subsidiary of XYZ Company?
B. XYZ Company could also integrate Target Company into its existing corporate IT operations. Web-based customer services, inventory management, order processing, and other activities can be readily added to cut down the required working capital by 50 percent from its stand-alone values. There is, however, an increased IT service charge of $1 million for the first year, and this charge increases by 4 percent per year. Again, assuming a 10 percent discount rate, what is the maximum price XYZ should pay for Target Company under the integration scenario?

Table 3.20. Income statement of Target Company (2008–future)

Entries	2008 (thousands of dollars)	2009 (thousands of dollars)	2010 (thousands of dollars)	2011 (thousands of dollars)	2012 (thousands of dollars)	Growth to infinity (%)
Sales	61,000	63,440	65,978	68,617	71,361	4
Cost of goods sold	29,890	31,086	32,329	33,622	34,967	4
Depreciation	4,000	4,160	4,326	4,499	4,679	4
Selling and administrative	21,010	21,850	22,724	23,633	24,579	4
IT services	0	0	0	0	0	4
EBIT	6,100	6,344	6,598	6,862	7,136	4
Tax (35%)	2,135	2,220	2,309	2,402	2,498	4
EBIAT	3,965	4,124	4,289	4,460	4,638	4
Cash flow	7,965	8,284	8,615	8,960	9,318	4
Additional investments	—	—	—	—	—	—
Working capital	8,000	8,320	8,653	8,999	9,359	4

Answer 3.11

For estimating the present values of cash flows for the period 2004 to infinity, the computations illustrated in Table 3.21 are needed. In Table 3.21, the letter A denotes a constant but unknown cash flow defined at the beginning of year 2012. The applicable derivation is shown below.

The present value of cash flow (at the beginning of year 2008) in Table 3.21 consists of the following parts:

1. 7,965/1.1
2. 8,284/1.1^2
3. 8,615/1.1^3
4. 8,960/1.1^4
5. 9,318/1.1^5
6. The present value of the cash flow amounts of 2013 to infinity (see equation in Table 3.21):

$$A = 9{,}318$$
$$\text{Formula} = A\ [(1 - r)/r)]/1.10\text{^}5$$
$$\text{Present value} = 9318 \times 17.33333/1.10\text{^}5 = 100{,}286.05$$

The sum of the above six items is 32465.31 + 100,285.05 = 132,750.45. This is what is correctly shown in Table 3.22.

Note: The closed-form equation $(1 - r)/r$ may be derived as follows:

$$C = (1 - r) + (1 - r)\text{^}2 + (1 - r)\text{^}3 + (1 - r)\text{^}4 + \ldots + (1 - r)\text{^}n$$
$$C\,(1 - r) = (1 - r)\text{^}2 + (1 - r)\text{^}3 + (1 - r)\text{^}4 + \ldots + (1 - r)\text{^}n + (1 - r)\text{^}(n + 1)$$
$$C - C(1 - r) = (1 - r) - (1 - r)\text{^}(n + 1)$$

$C = (1 - r)/r$ as n approaches infinity.

The same factor $(1 - r)/r$ is applicable for the estimation of the present values of working capital for the period of 2013 to infinity. The spreadsheet represented by Table 3.22 enumerates the results.

The NPV of the Target Company, as a stand-alone operation, is negative, not justifying its potential acquisition by XYZ Company. On the other hand, if XYZ integrates the Target Company into its IT operations, then the value of this Target Company is significantly improved, as displayed in Table 3.23.

Under the integration scenario, the Target Company is worth about \$116 million. Any bid price paid below this figure will represent a net gain for XYZ Company.

Table 3.21. Present value of cash flows in 2004 (growth rate = 4% and discount rate = 10%)

Entries	2012	2013	2014	2015	Infinity
Cash flow	*A*	*A*(1.04)	*A* (1.04)^2	*A* (1.04)^3	
Present value (2013)		*A* (1.04)/(1.10)	*A* (1.04)^2/(1.10)^2	*A* (1 – 0.4)^3/(1.10)^3	
		A (1 – *r*)	*A* (1 - *r*)^2	*A* (1 – *r*)^3	
PV of all future cash flow at start of 2008		*A* [(1 – *r*)/*r*] /(1.10)^5			
r	0.0545455				
(1 – *r*)/*r*	17.3333				

Table 3.22.. Income statement of Target Company (stand-alone subsidiary)

Entries	2008 (thousands of dollars)	2009 (thousands of dollars)	2010 (thousands of dollars)	2011 (thousands of dollars)	2012 (thousands of dollars)	Growth to infinity (%)
Sales	61,000	63,440	65,978	68,617	71,361	4
Cost of goods sold	29,890	31,086	32,329	33,622	34,967	4
Depreciation	4,000	4,160	4,326	4,499	4,679	4
Selling and administrative	21,010	21,850	22,724	23,633	24,579	4
IT services	0	0	0	0	0	4
EBIT	6,100	6,344	6,598	6,862	7,136	4
Tax (35%)	2,135	2,220	2,309	2,402	2,498	4
EBIAT	3,965	4,124	4,289	4,460	4,638	4
Cash flow	7,965	8,284	8,615	8,960	9,318	4
PV (cash flows)	132,750					4
Additional investments						
Working capital	8,000	8,320	8,653	8,999	9,359	4
PV (WC charge)	13,333					
(A) NPV of Target Company	−119,417					
	Not to acquire					

Table 3.23. Income statement of Target Company (stand-alone subsidiary)

Entries	2008 (thousands of dollars)	2009 (thousands of dollars)	2010 (thousands of dollars)	2011 (thousands of dollars)	2012 (thousands of dollars)	Growth to infinity (%)
Sales	61,000	63,440	65,978	68,617	71,361	4
Cost of goods sold	29,890	31,086	32,329	33,622	34,967	4
Depreciation	4,000	4,160	4,326	4,499	4,679	4
Selling and admin.	21,010	21,850	22,724	23,633	24,579	4
IT services	0	0	0	0	0	4
EBIT	5,100	5,304	5,516	5,737	5,966	4
Tax (35%)	1,785	1,856	1,931	2,008	2,088	4
EBIAT	3,315	3,448	3,586	3,729	3,878	4
Cash flow	7,315	7,608	7,912	8,228	8,558	4
PV (cash flows)	121,934					4
Additional investments						
Working capital	4,000	4,160	4,326	4,499	4,679	4
PV (WC charge)	6,667					
(A) NPV of integrated operations	115,268					
	To bid no more than $116 million					

3.7 CONCLUSIONS

This chapter introduces the basic accrual principle of financial accounting, discusses the workings of income statements, balance sheets, and funds flow statements, with explanations of all applicable accounting terms. Also pointed out are the ways in which contributions by engineering managers and professionals are recorded in these financial statements.

Ratio analysis uses the financial data contained in these statements to assess companies' financial health. EVA is described as an upgraded method of reporting the true financial value created by a company, unit, or a specific project.

The shortcomings of ratio analysis are outlined and a broad-based system of measurement metrics (the balanced scorecard) is illustrated. An adoption of this broad-based metrics system by corporate America will likely shift corporate management's attention from being predominantly focused on short-term financial results to a balanced emphasis on corporate long-term growth. This is possible with the use of metrics that address important competitive factors such as customer satisfaction, continuous improvement of internal business processes, and innovation for growth. As such broad-based measurement metrics become widespread, the critical contributions made by engineering managers and prodfessionals will be increasingly recognized and rewarded.

There are different types of capital projects in which a company might invest: operations (assets in place), opportunities (R&D and marketing), acquisitions, and joint ventures.

Different evaluation methods are used for these capital projects. For example, DCF (based on WACC), IRR, multipliers, and Monte Carlo simulations are useful for evaluating operations. Option pricing is suitable for evaluating opportunities for which there are no predictable cash flows. Acquisitions and joint ventures are advanced financial topics, the evaluation of which should be deferred to knowledgeable financial specialists on the subject.

Also briefly presented in Appendices is the practice of T-Accounts, which are used to document financial transactions. The concept of risks is also explained.

Arthur Ashe said: "One important key to success is self-confidence. An important key to self-confidence is preparation." Engineering managers and professionals are advised to become well prepared to actively participate in the company's evaluation criteria adopted for capital budgeting. Doing so will enable them to constantly bring forth and screen useful projects and valuable opportunities (including risk assessment) and to be in a position to initiate winning capital project proposals on a timely basis.

3.8 APPENDICES

3.8.1 T-ACCOUNTS

Accountants use T-accounts as tools to document transactions in preparation for creating financial statements. T-accounts are set up for any items that are assets, liabilities, equities, or other temporary holding entries. A T-account, as displayed in Figure 3.1, looks like the letter "T." On the left side of the "T," debits are recorded and on the right side credits are recorded. This type of entry is also known as double-entry book keeping. Figrue 3.2 shows some additional details of such T-accounts.

Following the double-entry bookkeeping practice, every transaction affects at least two accounts. This is to ensure that a balance is continuously maintained between both the assets of, and the claims against, the company.

The company's assets include cash, accounts receivable, inventory, land, machines, plant facilities, marketable securities, and other resources of value. Liabilities include accounts payable, accrued expenses, long-term debts, and other claims creditors have against the company. Owners' equities include stocks, capital surplus, retained earnings, and other

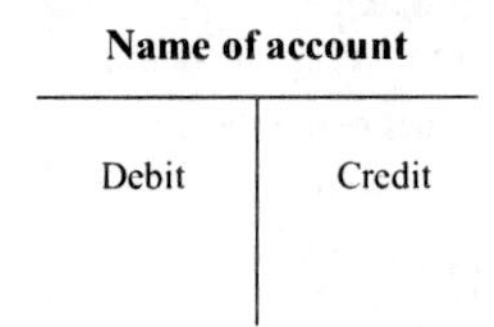

Figure 3.1. T-account.

STORES		WIP		FG	
Debit	Credit	Debit	Credit	Debit	Credit
(1)	(2)	(3)	(4)	(5)	(6)

Explanations:
(1) Purchasing raw materials
(2) Putting materials into production process
(3) Production is initiated, adding value to raw materials.
(4) Production is complete
(5) Receiving of finished goods in storage
(6) Finished goods are shipped for sale

Figure 3.2. T-Accounts with additional details.

claims of the owners against the company. Equation 3.9 depicts the balance between assets (A) and claims consisting of liabilities (L) and OE:

$$A = L + OE \tag{3.9}$$

The convention of T-accounts is as follows: To increase the amount of an asset, debit the account; conversely, to decrease an asset amount, credit the account. All liabilities and OE accounts are treated in the opposite way; that is, to increase a liability or equity amount, credit the account; and to lessen a liability or equity amount, debit the account.

For accounts that do not fall directly into one of these three categories (i.e., A = assets, L = liabilities, and OE = Owners' equities), we need to first define their relationships to either A, L, or OE and then treat them accordingly. Revenues, expenses, and dividends are such examples.

Revenues raise the net income of the company. The resulting net income goes into the retained earning account for the owners. Thus, an increase of revenues needs to be credited to its T-account. The company's expenses are generally deducted from its revenues in order to arrive at its net income. An increase in expenses results in a reduction of net income and consequently a reduction of OE. Therefore, an increase of expenses needs to be debited to its T-Account. Similarly, an increase in dividends paid to shareholders whittles down the residual net income amount, which is then added to the retained earnings account of the owners. Thus, an increase of dividends needs to be debited to its T-Account. Table 3.24 summarizes the ways in which increases in the indicated assets, liabilities, OE, or other amounts should be treated in their respective T-accounts.

For engineers and engineering managers who are familiar with equations, the rule that follows may represent a convenient way of keeping them better oriented with the T-account convention. Rearranging the basic accounting equation (Equation 3.9), we get

$$\text{LHS} = A - L - \text{OE} = 0 \tag{3.10}$$

where LHS stands for "left-hand side" of the equal sign. Note that, for each financial transaction, there are two account entries. The account entry that causes the LHS to increase temporarily should be debited to its respective T-account. Examples include increases in all assets and decreases in all liabilities and OE. The account entry, which leads to a temporary reduction of the LHS, should be credited to its respective T-account. Equation 3.10 remains valid after both entries of the financial transaction are entered.

Table 3.24. T-Account convention for an increase in selected account items

Accounting items	Debit	Credit
Assets	x	
Cash, accounts receivables, inventory, land, machines, marketable securities, etc.		
Liabilities		x
Accounts payables, accrued expenses, long-term debts, etc.		
Owners' equities		x
Stocks, capital surplus, retained earnings		
Revenue		x
Expenses	x	
Dividend	x	

Source: David F. Hawkins and Jacob Cohen, "The Mechanics of Financial Accounting." *Harvard Business School Note*, No. 9-101-119, June 27, 2001.

Accountants use *T-accounts to collect raw financial* data that they check and recheck for validity and reliability. Then they make sure that the data are relevant to the accounting period at hand and consistent with past practices. Finally, to ensure comparability, accountants regroup them into known line items typically included in financial statements.

Example 3.12

Study the following accounts, which contain several transactions keyed together with letters in Table 3.25. Explain the nature of each transaction with the dollar amount involved.

Answer 3.12

(a) Convert capital of $9,000, add $6,000 to the cash account, and purchase books worth $3,000 for the Law Library.
(b) Pay the prepaid rent of $2,500 in cash.

Table 3.25. Various T-accounts

Cash		Office Equipment		Capital	
(a) 6,000	(b) 2,500	(d) 8,500			(a) 9,000
(e) 1300	(c) 150				
	(f) 3,500				
	(g) 160				

Office Supply		Law Library		Legal Fees Earned	
(c)150		(a) 3,000			(e) 1,300
(d)125					

Prepaid Rent		Account Payable		Utility Expenses	
(b) 2,500		(f) 3,500	(d) 8.625	(g) 160	

(c) Pay office supplies of $150 in cash.
(d) Purchase office equipment ($8,500) and office supplies ($125) by credit, creating an account payable of $8,625.
(e) Receive the $1,300 legal fees earned in cash.
(f) Pay accounts payable of $3,500 in cash.
(g) Pay utilities expenses of $160 in cash.

3.8.2 RISKS

In general, the outcome (i.e., earnings) of any investment has a degree of inherent uncertainty: large in some and small in others. Investment risk is defined as the measure of potential variability of earning from its expected value. It is usually modeled mathematically by the standard deviation of the outcome when the outcome is expressed in the form of a probability density distribution function. For common stocks, risk is modeled by a relative volatility index, Beta, as defined in Figure 3.3.

The rate of return on risky security can be modeled as the risk-free rate plus a risk premium:

$$r = R_f + R_p \tag{3.11}$$

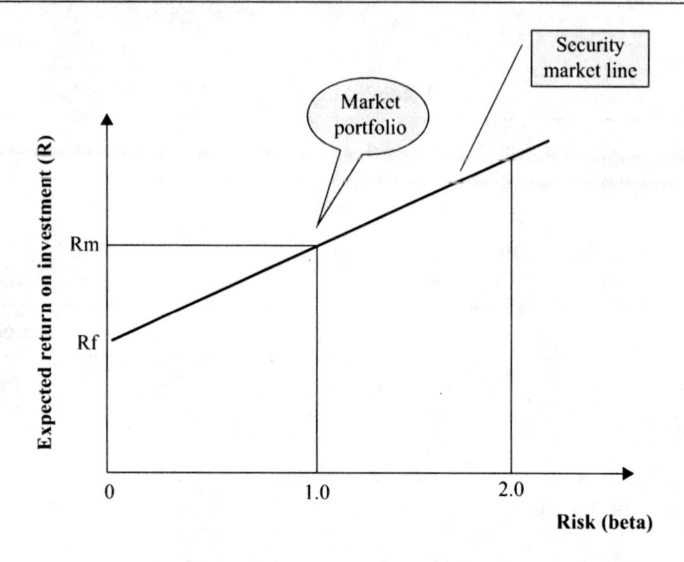

Figure 3.3. Security market line

The risk-free rate (e.g., R_f equals some constant rate such as 4.0 percent for 10-year U.S. Treasury bills) is the return for compensation of opportunity cost without uncertainties—assuming that the U.S. government will never defaut in discharging its debt-paying responsibility. The risk premium is the additional return needed to compensate for the added risks the investors undertake.

Figure 3.3 displays the general risk–reward correlation between expected annual return and the associated risk of the investment in question. An investor may realize only 4 percent from the risk-free U.S. Treasury bills, but a whopping 25 percent from highly precarious junk bonds.

The market portifolio is defined by the 500 stocks in the Standard & Poor's index, and it is assumed to model the average behavior of the stock market. For this market portifolio, the risk factor beta (or volatility) is define as "1." Investments that has higher risks will thus have a beta value larger than 1, and those with less risk are expected to have a beta value less than 1. Investments having high beta values are expected to receive higher returns.

Expected value is the return of an outcome multiplied by its probability of occurrence. For a portfolio containing N independent investments, the total expected return is the sum of the products of individual returns and its respective probabilities of occurrence. The total expected return is the average return weighted by probability factors:

$$\text{EV} = [\sum P_i \times R_i;\ i = 1 \text{ to } N] \quad (3.12)$$

where:

R_i = return of investment i

P_i = single – valued probability of occurrence for R_i

$[\sum P_i;\ i = 1 \text{ to } N] = 1.$

Risk aversion delineates the generally expected unwillingness on the part of investors to assume risks without realizing the commensurate benefits. Most investors are unwilling to assume risks unless there are incremental benefits (the risk premium) that compensate them for bearing the added risks involved. Some audacious investors are more willing than others to take on additional risks in order to realize the added benefits.

Table 3.26 exhibits an example in which the behavior of risk-averse investors can be studied. Investment A has a nominal value of $100 when the economy is in a normal state. This investment is projected to be valued at $90 if the economy goes into a recession in the near future. On the other hand, its value may increase to $110 if the economy booms. Investment B also has a nominal value of $100 when the economy is normal, but its value decreases to zero in a recession and increases to $200 in a booming economy.

If we further assume that there is an equal probability of 33.33 percent that the state of the future economy will be either normal, in recession, or in a booming state, then the expected pay-off value of these two investments is identical (e.g., $100); however, their amounts at risk are different: –$10 for investment A and –$100 for investment B.

Table 3.26. Behavior of investors

State of economy	Investment A (dollars)	Investment B (dollars)	Probability
Recession	90	0	0.333
normal	100	100	0.333
Boom	110	200	0.333
Expected pay-off value	100	100	
Range	90–110	0–200	
Amount at risk	–10	–100	

Note: Conservative persons would choose investment A, whereas risk-preferring persons would choose investment B. Risk-neutral persons would not have a preference.

Among these two investment options, a risk-averse investor will choose investment A for its lower downside risk; he or she will not choose investment B because there is no gain in return for the added risks. A risk-preferring investor will choose investment B for its reward potential of doubling the money if the economy booms in the future.

3.8.3 RISK MANAGEMENT

Companies need to consider a variety of business risks and develop strategies to timely address them. The following five types of risk are common [Walker[35]]:

1. Regulatory risks: Changes in law, tariffs, taxes and other public policies.
2. Reputational risks: Impact of brand and corporate image due to news, rumors and stories, as related to company's products/services, labor practices, environmental impacts, or supplier relationships.
3. Operational risks: Due to inadequate or failed internal processes, people, and systems, or from external events.
4. Credit risks: Probability of not being timely paid on Accounts Receivables.
5. Market risks: Changes in market conditions, such as customer's needs, competition, and advancement of technologies.

Companies develop a number of best-practice strategies to mitigate risks. Example of such risk management strategies would include:

(a) Collect verifiable risk information diligently, as a part of their company's knowledge management practice.
(b) Evaluate the propensity of risks the company is exposed to in the above-described categories for a reasonably long time frame of reference (e.g., 5 years).
(c) Benchmark own risk management practices with other companies to remain among the leaders in the industry.
(d) Pay special attention to operational risks, as "the devil is in the details."
(e) Make team-based decision to mitigate risks, relying primarily on data and on verifiable facts.
(f) Keep record of lessons learned for the company to become better in risk management over time.

CHAPTER 4

Marketing Management

4.1 INTRODUCTION

"Business has only two basic functions—marketing and innovation," says Peter Drucker. Marketing develops and supports the enterprise's efficacy of competition by offering better products/services (e.g., having richer features, higher reliability, being more customizable, selling at lower costs, etc.) than its rivals. Marketing impacts the top line (i.e., sales revenue) of the company.[1] Innovation strengthens the enterprise's position to sustain profitability by applying unique technologies and other core competencies. As the markets further proliferate due to explosions of customer segments, service technologies, communication methods, and distribution channels, the marketing function is becoming increasingly complex, more costly, and less effective today than before.[2]

Product/service marketing is the general subject of a number of relatively new textbooks.[3] Others cover certain specific topics, such as customer loyalty[4] and relationship marketing.[5] Still others address the marketing details in specific sectors, such as professional services,[6] architectural services,[7] financial services,[8] legal services,[9] telecommunication services,[10] hospitality marketing,[11] and health care marketing.[12]

The objective of this chapter is to introduce various basic marketing concepts related to products/services[13] and to prepare engineering managers and professionals to interact effectively with marketing and sales personnel in for-profit companies. Also presented are concepts and applications regarding marketing management processes, identification of opportunities and threats facing an organization, and marketing tasks.

Engineering managers and professionals are known to be technologically innovative. We all know that engineers tend to have a natural proclivity for technological innovation. By mastering marketing techniques as well, they can make significant contributions toward the success of their companies and as a result, possibly gain entry into senior leadership positions.They should be prepared to continue studying additional current references on specific issues in marketing management.

4.2 THE FUNCTION OF MARKETING

Marketing and sales are critically important functions to profit-seeking companies because they strive to ensure satisfaction in the exchange of values between the producers and consumers of products/services, as exhibited in Figure 4.1.

4.2.1 SALES VERSUS MARKETING

Sales is a process by which producers attempt to motivate target customers to buy the available products/services. The mentality behind sales, as illustrated by Figure 4.2, is that "someone out there will need this." At the time when Ford Motor Company was the dominant carmaker in Detroit, the well-known saying attributed to Henry Ford was, "You can have any

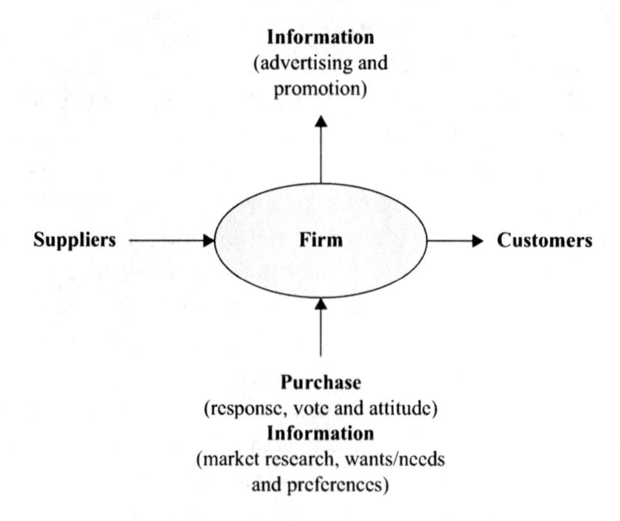

Figure 4.1. Marketing interaction.

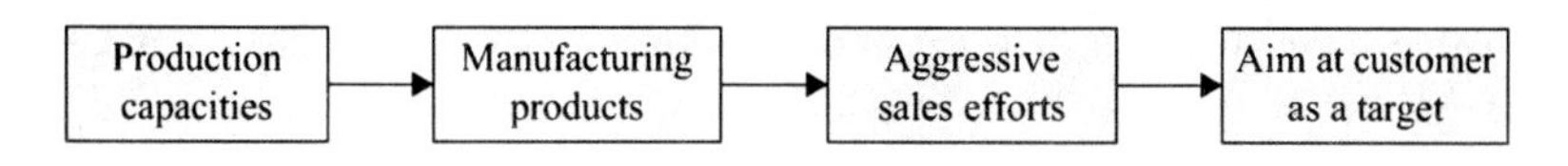

Figure 4.2. Sales orientation

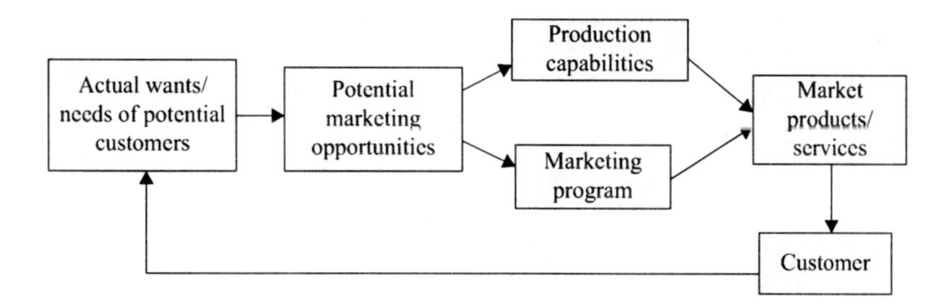

Figure 4.3. Marketing orientation.

color you want for your car, as long as it is black." Sales does not take the customers' concerns into account.

In contrast, companies with a marketing orientation offer something customers want by seeking feedback from the marketplace, adjusting the product/service offerings, and enhancing the value to consumers (see Figure 4.3).

In the pursuit of marketing strategies, companies solicit intelligence, financial data, and customers' responses to constantly reassess the market. They evaluate such factors as changing needs, competition, cost effectiveness, product/service substitution, and maturity of services. A long-term orientation ensures that benefits for both producers and customers will be sustainable. Sales strategies are only a part of marketing.

4.2.2 THE MARKETING PROCESS

The marketing efforts of companies are typically focused on four specific aspects:[14] (a) **customer focus:** The purpose of a profit-seeking business is to understand customers' needs, deliver value to them, and offer services to ensure that they are satisfied. In other words, the customer comes first. (b) **competitor focus:** Companies seek advantages relative to competitors, monitor competitive behaviors, and respond to the strategic moves of competitors. (c) **inter-functional coordination:** Companies integrate all functions, share information, and organize themselves to provide added value for the customers. (d) **profit orientation:** Companies focus on making profits in both short term and long term.

To achieve business success, companies must search constantly for future markets, in addition to actively serving the markets of today. The primary emphasis of marketing is to scan the relevant business environment for future opportunities (such as what bundle of products and services to offer to whom, at what price, at what time, and in which market segments) and to provide insight into the needs of current customers and the intentions of competitors.

Presented in Figure 4.4 is the marketing process, which is iterative in nature. This process defines (a) opportunities (unsatisfied needs) in the marketplace; (b) the products or services with the proper features to satisfy these needs; and (c) the product/service pricing, distribution, and communication strategies to serve the target market segments. Market segments refer to specific groups of customers who share similar purchasing preferences.

For companies to succeed in the marketplace, marketing must be recognized as a core activity, central to the company's strategy formulation and execution. Through marketing, companies identify and satisfy the needs of customers and achieve long-term profitability by attracting and retaining these customers. Specific tasks undertaken to attain these objectives include (a) interacting with and understanding the market and its customers, (b) planning long-term marketing strategies, and (c) implementing short-term tactical marketing programs.

The effectiveness of a marketing program is often measured by two metrics: (a) how attractive are the company's products/services to the target customers and (b) how successfully can the company satisfy and retain these target customers. Figure 4.5 illustrates the marketing effectiveness diagram.

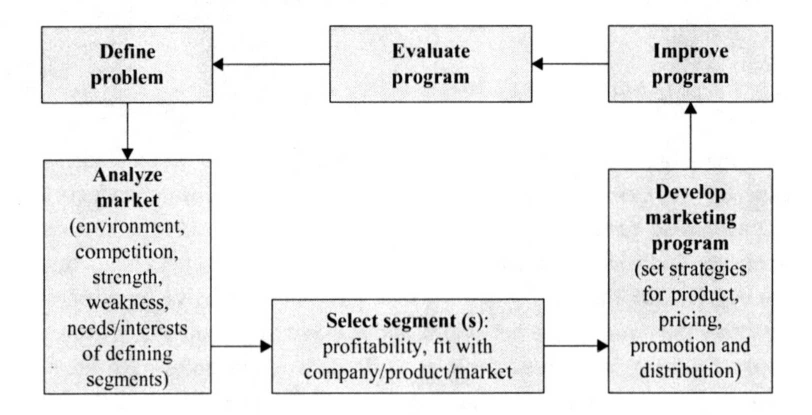

Figure 4.4. Marketing process.

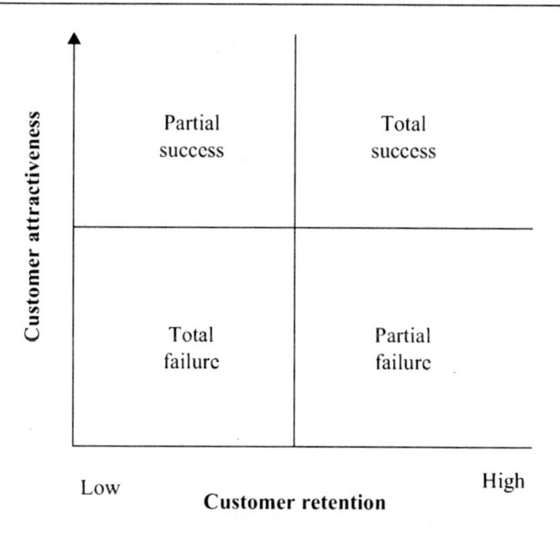

Figure 4.5. Marketing effectiveness diagram.

The marketing program of a company is to be regarded as a *total success* if both the customer retention and attractiveness of products/services to customers are high. This is when high profitability can be achieved at a maximum growth rate. The marketing program represents a *partial success* if the product/service attractiveness to the customer is high, but the customer retention is low. While lost customers are typically replaced by new customers, the total customer base will show little growth. *Partial failure* of a marketing program is recognized when product/service attractiveness to the customer is low, but the customer retention is high. In this scenario, business remains stagnant because it relies on loyal customers to repeatedly buy mature and noncompetitive products/services. The company's sales may slow down or fall as few new customers are added. The marketing program is said to be a *total failure* if both product/service attractiveness to the customer and the customer retention level are low. Customers leave, and the company's sales fall.

Marketing strategies are implemented at the corporate, business, and operational levels. Top management provides inputs to identify future opportunities, and addresses such questions as what business the company is in and what business the company should be in. The business managers then specify their ideas, bring out products or services, and strive to create and maintain a sustainable competitive advantage in the marketplace. At the operational level, managers and support personnel conduct the planning for specific marketing programs, and implement and control marketing efforts related to segmentation, product (service), pricing, distribution channels, and communications.

4.2.3 PRODUCT AND SERVICE MARKETING

Companies offer value propositions to the marketplace in the forms of products or services, or both. Products represent tangible physical entities (e.g., automobiles, jet engines, iPhones, iPad Air, Microsoft Surface 2, etc.), which embody specific features and functionalities to deliver value to their end users. Usually, the design, production, and consumption of products take place at separate times and locations, even though the consumer's feedback is aggressively solicited as inputs to the design process. Services, on the other hand, are intangible knowledge-intensive bundles of value aiming at causing a transformation of the end user's state in (a) knowledge (business consulting, education, information search, operations); (b) wealth (financial advisement, tax services); (c) health (physicians and hospital services, insurance, food services); (d) enjoyment (video rentals, entertainment services); (e) physical movement (e.g., car rentals, travel, relocation, transportation services); and others. In general, the design, production, and consumption of services occur at the same time and location and consumers participate actively in the whole process. Other companies deliver product-enhancing services to support the utilization of their products (e.g., maintenance, problem-solving, and operations). In general, services differ from products in the following four ways.

A. Lifecycle

Consumers who purchase a product, gain ownership over the product, and use it for as long as they want during the product's life cycle. Consumers who purchase a specific service will obtain only the rights to use it during a specific period of time (e.g., services related to financial investing, banking, legal advisement, medical consultation, education, and transportation). Because the life cycle of services is usually shorter than that of products, consumers tend to be more sensitive to the quality of services than to the quality of products.

B. Customer Experience

Products are tangible in that they possess measurable physical features, such as size, weight, and color, whereas services are considered intangible, even though both products and services produce long-lasting values that are useful to consumers. Customer experience derived from a product is typically affected by its primary value proposition (e.g., price, product features, product quality, frequency of use, and support services). Customer experience derived from a service is affected by a different set of value propositions (e.g., price, importance of the change of state made possible by the use of service, attitude of vendor's service staff, service quality, physical environment in which service is delivered and consumed, etc.). Customer satisfaction is harder to accomplish with services, as customers' expectations may be subjectively divergent due to an increased level of person-to-person interactions in the delivery and consumption phases of services.

C. Time and Place of Production and Consumption

Products are usually produced at a factory location, then transported to another warehouse location, and further displayed at a retail store location, to be purchased and used by customers at yet another entirely different time and location. Services, on the other hand, are typically produced and consumed at the same time and at the same place (e.g., financial advisement, seeing a show, and visiting a physician). Thus, services cannot be separated from the service providers. Products may be stored, whereas services may not. Services last for a specific time and must be consumed at the same time as it is produced. Again because of this time constraint consumers demand more.

D. Variability in Products/Service Quality

The quality of products is usually quite uniform, if advanced quality control means are properly applied during the manufacturing process. The quality of service depends, to a large extent, on the interpersonal skills, attitude, service knowledge, and the support process in place to resolve customers' problems. It is very difficult for service providers to make all customer service experiences identical.

Because of these differences, the marketing of services needs to be more attentive to factors such as physical evidence, process, and people, as compared to the marketing of products.

4.2.4 KEY ELEMENTS IN MARKETING

Workers who plan and implement marketing programs are called *marketers*. Marketers consider various influence factors and make diverse decisions to penetrate a specific marketplace.[15] They pay attention to several key elements of marketing, such as the market itself, the environment, the customer, the product (service), pricing, promotion, and distribution.

The ***market*** is made of buyers who are expected to purchase certain products/services, and also buy substitutes that offer similar values. Of great importance to the marketers is the size of the market, measured in millions of dollars per year; its growth rate; its location characteristics; and its requirements. The market must be large, stable enough with a reasonable growth rate, and relatively easy to reach and serve in order for it to be an attractive target for the marketers. One would have to be under a rock to not know that the market is global. Nowadays, having a global orientation is no longer opulence, but a necessity for economic survival in any industry.

The ***environment*** refers to competitors, barriers to entry, rules and regulations, resources, and other such factors affecting the marketers' success in a given market segment. Marketers must also understand the opportunities and threats present in the environment. They cannot be oblivious to the changing in customer perspectives of products/services in the marketplace. As the boundaries among countries, industries, and market segments cumber, every enterprise is facing a plethora of competition.

The ***customers*** consist of all potential buyers of a given product/service. Companies need to understand (a) why they buy, (b) how they buy, (c) who makes the decision to buy, (d) who will use the product/service in question, (e) in what specific way the use of the product/service will contribute value to the user, (f) what might be the buyer's preference related to support service and warranty, and (g) what other product/service features the customer may want, as well as other factors. Appendix 4.9.1 contains additional sample customer-survey questions. It is of course important for marketers to recognize that the customers are often their most valuable assets and they need to do everything under their control to attract and keep them. A company that uses marketing strategies to understand its customers' needs can then implement service strategies that build and maintain competitive advantages in the marketplace. Beware of the instances wherein customers may not always know what they want, such as innovative and sight-unseen products/services.

To become customer-oriented, companies need to (a) define the generic needs of customers through research (such as the buyer's perception of the cost, status, and value of a financial advisement service); (b) identify the target customers by segmentation (including which selected groups of customers have shared needs); (c) differentiate products/services and communication channels (e.g., offering special reasons for customers to buy through unique product/service attributes or unique communications); and (d) bring about differentiated values for customers.

Consumers and business customers have different service needs. Products/services are not of the same value to all their intended end users, nor do they require the same depth of knowledge and technology to deliver. Because rational end users are willing to pay proportionally more for high-value products/services, marketing efforts must be devoted to products/services in proportion to their value to the end users. Implementing such a value-based marketing strategy will have a direct impact on corporate profitability derivable from the marketplace. Figure 4.6 shows some service examples based on value and the extent of knowledge and technology needed to deliver them.

For services in each of these six categories shown in Figure 4.6, companies will need to adopt suitable marketing strategies to optimally promote awareness, build brand image to attract their targeted consumers and business customers, effect their buying decisions, and retain them.

Example 4.1

To us engineers and technologists, products and services are the key "bundles of value" we create as the primary basis for any enterprise to become profitable over time. In the absence of such "bundles of values" the enterprise has nothing meaningful to offer to the marketplace. Things we do

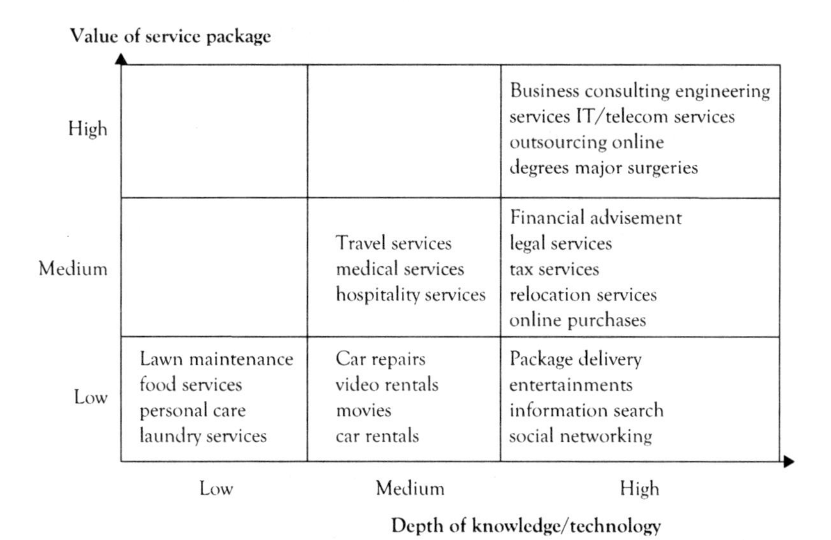

Figure 4.6. Service value versus knowledge/technology.

and say are usually reproducible and verifiable. We are clearly the backbones of any enterprise.

Please explain why any enterprise needs a marketing team? Why should we engineers/technologists bother wasting our time to collaborate with these marketing guys who talk a funny language, use imprecise concepts, invoke data they can hardly prove, wave their hands, and project such an air of optimism that can seem rather arrogant at times? Why should we engineers/technologists respect this "fast-talking bunch" and treat them as our "equals" in the first place?

Answer 4.1

For any product/service enterprise to succeed in the market place, it needs innovative products/services that customers are willing to buy. There are indeed two aspects in this business model. Engineers and technologists are important contributors to the development of innovative products/services, which are low cost, reliable, and easy to deliver and support. This core product/service element is the key part of any business venture.

On the other hand, without marketing, engineering would not know which product/service features to emphasize that would satisfy the needs of customers while offering meaningful differentiations to the competition. Furthermore, marketing creates customer awareness through the right advertising channels; such awareness is critical to achieve commercial success by any product/service. Marketing also (a) divides the customer base into segments so that they can be better served, (b) communicates with customers constantly to understand their future needs, (c) manages the customer relationships to discover unique usage patterns that might be important for the development of new services, and (d) tracks the competition to mitigate potential threats.

Successful product/service enterprises need to demand that marketing and engineering are collaborating effectively in order to assure sustainable profitability.

4.3 MARKET FORECAST—FOUR-STEP PROCESS

The purpose of conducting a market forecast is to define the characteristics of the target market as to, for example, size, stability, growth rate, and serviceability. Market size and growth rate must be large enough to warrant further consideration by marketers. Any future market is always uncertain due to potential changes in end-user behavior, global economics, new technologies, competition, and economic and political conditions. The key to successfully forecast market size is to understand the underlying forces behind the demand.[16] Barnett proposes a four-step process, as follows: (a) define the market, (b) segment the market, (c) determine the segment drivers and model their changes, and (d) conduct a sensitivity analysis. These steps are explained below.

4.3.1 DEFINE THE MARKET

On the basis of customer interviews, the market should be defined broadly to include the principal product/service to be marketed, its "bundle of benefits" to customers, and product/service substitution.

4.3.2 SEGMENT THE MARKET

In segmentation, the potential customers for the principal product/service are divided into homogeneous subgroups (segments) whose members have similar product/service preferences and buying behavior. Market segmentation is elucidated in detail in Section 4.4.

4.3.3 DETERMINE THE SEGMENT DRIVERS AND MODEL THEIR CHANGES

Segment drivers may be composed of macroeconomic factors, such as the increase in white-collar workers and changes in population, as well as industry-specific factors, such as the industrial growth rate and business climate. Possible sources of information related to segment drivers are industrial associations, governments, industrial experts, marketing data and service providers, and specialized market studies.

4.3.4 CONDUCT A SENSITIVITY ANALYSIS

Sensitivity analyses are conducted to test assumptions. Monte Carlo simulations may be performed to generate the maximum—most likely—and minimum total market demand values, as well as an assessment of the risks involved (see Section 2.6).

The following are illustrative examples in which the total market demand for a product/service is estimated by (a) defining the industrial segments that purchased the product/service in the past, (b) determining the future growth rates of these industrial segments, and (c) calculating the total market demand for the product/service with these industrial segment growth rates as the segment drivers. The assumption here is that the demand of products/servcies in a given industrial segment is in direct proportion to its segment growth rate.

For example, to predict the demand of electricity in future years, a utility company may subdivide its consumers into three segments: industrial, commercial, and residential. The need for electricity by the industrial segment depends on its future production level and business climate. The electricity demand by the commercial segment is related to retail sales that in turn are negatively affected by retail stores consolidating and by growing Internet sales. For example, web-based sales increased by 28.5 percent to $14.8 billion in 2002. The residential electricity demand is affected positively by new home sales and home sizes and negatively by the increased energy efficiency of home appliances.

A second example is a paper-producing company that has determined the total market demand for uncoated white paper by deconstructing the market into end-use segments such as business forms, commercial printing, reprographics, envelopes, stationery, tablets, and books. The drivers in each segment are then modeled in terms of macroeconomic and industrial factors, using regression analysis and statistics. Examples of applicable drivers include growth in the use of electronic technology, white-collar workers, the present level of economic activity, the growing use of personal printers, population growth, demand growth induced by price reduction, and the practice of paying bills online and not by checks stuffed into envelopes.

Market forecast is a difficult, but critical first step to take when developing a marketing program. Companies routinely engage both internal and external resources to assess the principal characteristics of the target market for their products/services.

4.4 MARKET SEGMENTATION

Once it is determined that the target market is worth pursuing (i.e., the market size is large and stable enough with a high growth rate), then it is useful to understand the potential customers in such a market so that they can be served well. Market segmentation is a process whereby companies recognize the differences between various customer groups and define the representative group behaviors. Members in each of these customer groups respond to product/service offerings in similar manners and have comparable preferences with respect to the price–quality ratio, reliability, and service requirements.

4.4.1 PURPOSE OF MARKET SEGMENTATION

By dividing consumers into groups that have similar preferences and behaviors with respect to the products/services being marketed, companies achieve four specific useful purposes: (a) matching products/services to appropriate customer groups, (b) creating suitable channels of distribution to reach each of these customer groups, (c) uncovering new customer groups that may not have been served sufficiently, and (d) focusing on niches that have been neglected by competitors.

Overall, segmentation allows companies to realize the following benefits: (a) developing applicable marketing strategies and objectives, (b) formulating and implementing marketing programs that address the needs of the different customer groups, (c) tracking changes in buying behavior over time, (d) understanding the enterprise's competitive position in the marketplace, and recognizing future opportunities and threats, and (f) utilizing marketing resources effectively.

4.4.2 STEPS IN MARKETING SEGMENTATION

Displayed in Figure 4.7 is a segmentation flow diagram that illustrates the key steps in segmenting a market.

Companies need to classify consumers into segments by understanding their individual, institutional, and service-related characteristics. *Individual* characteristics include culture, demographics, location, socioeconomic factors, lifestyles, family life cycle, and personalities. *Institutional* characteristics include the type of business, its size, and extent of global reach. *Service-related* characteristics include type of user, original equipment manufacturer (OEM) versus private end user, usage level, service knowledge, brand preference, and brand loyalty. Also to be studied are benefits sought by consumers such as psychological and emotional benefits, functional performance, and price.

Millions of consumers purchase cars every year. To some, cars represent a status symbol; to others, cars are simply a means of transportation.

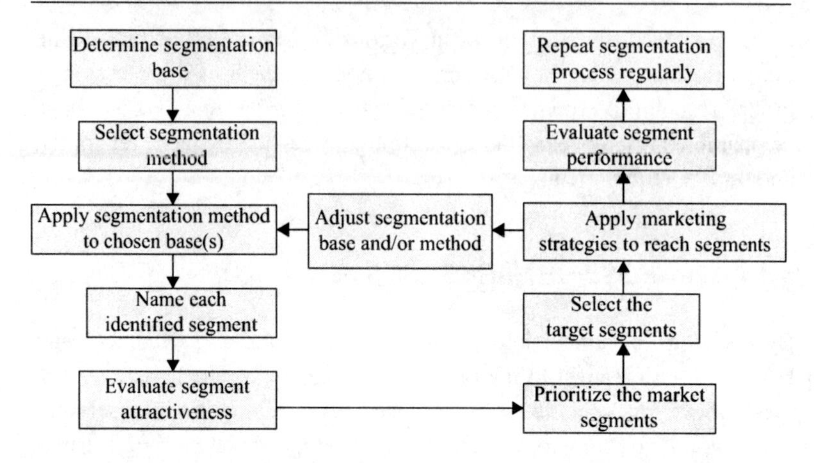

Figure 4.7. Segmentation flow diagram.

A large number of car buyers emphasize safety and reliability, while others focus on fuel economy. Socioeconomic factors, demographics, personalities, and family life are all known to influence the behavior of car buyers. These consumers are extensively segmented by all major carmakers.

4.4.3 CRITERIA FOR MARKET SEGMENTATION

To be effective, the segmentation of a market needs to satisfy several requirements. The segmentation should be measurable. It should result in readily identifiable customer groups. The identified customer groups should also be homogeneous. Each group's members should possess more or less unified value perceptions and display compatible behavioral patterns. These customer groups are reachable by promotion and distribution means. Above all, the segments should be large enough in size to justify marketing efforts, and they should have a high-growth rate to allow the company to achieve long-term profitability.

4.4.4 PITFALLS OF MARKET SEGMENTATION

There are pitfalls to market segmentation. Certain "old economy" companies adopt the asset-rich business strategy of pursuing the scale of economy advantages. For these companies, a potential pitfall is over-segmentation, because the selected segments may be too small or fragmented to serve effectually. Such an over-segmentation is not a pitfall for other "knowledge economy" companies that form partnerships to establish supply chains for "build-to-order" products/services. To foster differentiation, knowledge economy companies pursue product/service customization as the basis for their business strategies.[17] Examples of these products and services are mini-brewers, computer systems, custom cosmetics, financial advisement, and architectural design services. For other companies, overconcentration (lack of balance between segments) could have a negative impact on their overall marketing effectiveness.[18]

Market segmentation is a prerequisite to developing a workable marketing program. Knowledge derived from customer groups serves as valuable inputs to product/service design, pricing, advertising, and customer services, all of which are yet to be finalized by the marketers.

4.4.5 REDISCOVERY OF MARKET SEGMENTATION

Segmentation is designed to identify groups of customers who are particularly receptive to a particular brand of product/service, as well as being sufficiently large and lucrative to justify an active pursuit. Customer's traits such as values, tastes, and preferences are more likely to influence their purchase decisions than age, sex, educational background, and income level.[19] Segmentation must be done regularly, as customers' buying patterns (e.g., based on needs, attitudes, and behavior) may change over time and can be reshaped by market conditions (e.g., economics, new consumer niches, and new technologies). Effective segmentation focuses on one or two issues and they need to be reconsidered as soon as they have lost their relevance. Yankelovich and Meer[20] suggested a "Gravity of Decision Spectrum" described below to focus on the relationship of consumers to a product or service category:

1. **The shallow end:** Consumers seek products and services they think will save them time, effort, and money. Segmentation should focus on price sensitivity, habits, and impulsiveness of target consumers.
2. **In the middle:** Customers buy big ticket items (cars and electronics). Segmentation focuses on concerns about quality, design, complexity, and the status the product/service might confer.
3. **The deepest end:** Customers' emotional investment is great and their core values are engaged. These core values are often in conflict with market values. Segmentation needs to expose these tensions. Health care is an example service for these customers.

We have modified the original "What is in Stake Diagram"[21] for products/services as shown in Table 4.1.

Table 4.1. Modified "gravity of decision spectrum" for products/services

No.	Decision	Issues to address	Consumers' concerns	To find out by segmentation	Examples of services
1	Shallowest level	Whether to make small improvements to existing services	How relevant and believable the new service claims are	Buying and usage behavior	Information search (Google), car repairs, car rentals, video rentals, personal care (haircuts), entertainment, food services (Starbucks), package delivery
		How to select targets of a media campaign	How to evaluate a given service	Willingness to pay a small premium for higher quality	
		Whether to change prices	Whether to switch services	Degree of brand loyalty	
2	Middle level	How to position the brand	Whether to visit a clinic about a medical condition	Whether the consumers being studied are do-it-yourself or do-it-for-me types	Online degree programs, travel services, eBay purchases, physician services
		Which segments to pursue	Whether to switch one's brand of service	Consumer's need (better service, convenience, functionality)	

		Whether to change the service fundamentally	Whether to sign on an online degree program	Their social status, self-image, and lifestyle	
		Whether to develop an entirely new service			
3	Deepest level	Whether to revise the business model in response to powerful social and economical forces changing how people live their lives and how corporate business strategies are revised	Choosing a course of medical treatment	Core value and beliefs related to the buying decision	Consulting (business, outsourcing, financial advisement, legal), relocation, major medical services
			Deciding where to live		

Source: Adopted and modified from Yankelovich and Meer (2006).

Table 4.2. Marketing strategies for specific segments

Marketing variables	Segment 1	Segment 2	Segment 3
Price	Low	Low	High
Product/service	Standard	Quality	Quality and reliability
Promotion	Broad	Limited	Focused
Place	Multiple distributions	Multiple distributions	Direct
Physical evidence	Not important	Low emphasis	High emphasis
Process	Highly efficient	Efficient	Standard
People	Standard	Standard	Dedicated

Example 4.2

A company has divided the market for its existing services into three segments: (1) mass-market applications, (2) applications requiring a quality service, even though consumers continue buying on price, and (3) critical applications to which both quality and reliability are important. Advice the company on the marketing strategies that should be applied to these three segments.

Answer 4.2

For the company to be successful, a different set of marketing strategies needs to be applied to each of these segments, as suggested in Table 4.2.

Multiple distributions are recommended, including mass-merchandise department stores for wide distribution. Direct distribution should include catalogs, specialty stores, and upscale department stores.

4.5 MARKETING MIX (SEVEN PS)

The seven controllable variables in the marketing of services are price, promotion, product (service), placement (distribution), physical evidence, process design, and people; see Figure 4.8. The marketing mix for products consists of only the first four Ps. The last three Ps, namely physical evidence,

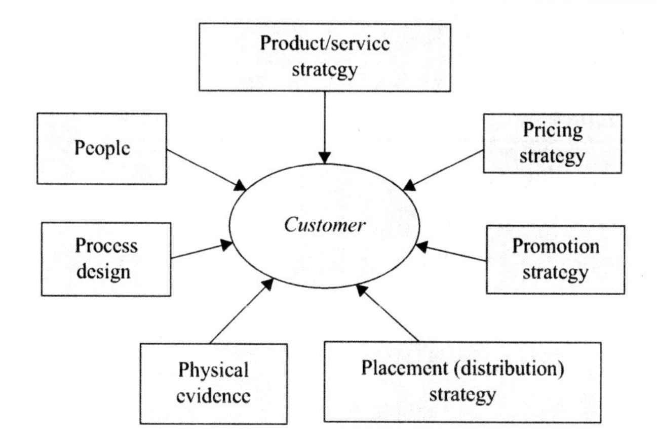

Figure 4.8. Marketing mix.

process design, and people, are added to account for the significant impact these factors exert on the marketing of services.[22] These seven variables characterize the multidimensional nature of marketing and are centered on customers, who are front and center to any successful marketing program.

4.5.1 PRODUCT (SERVICE) STRATEGY

The ***product/service*** symbolizes the actual "bundle of benefits" that is offered to customers by the marketers. Factors considered include functional attributes, appropriateness to customer needs, distinguished features over competition, product/service-line strategy, and product/service-to-market fit.

The product/service strategy takes the center stage in any marketing porgam.[23] If marketed properly, products and services that offer unique and valuable functional features to consumers are expected to enjoy a strong marketplace acceptance.

Products/services may be generally classified as either industry or consumer oriented. Their characteristics are different, as shown in Table 4.3. Marketing programs for consumer products/services are quite different from those for industrial products/services, even though the same basic marketing approach applies to both.[24]

A good marketing program must take into account the consumer's perception of products/services. Indeed, consumers perceive products/services in different ways from the producers and marketers. When buying products/services, consumers look for "bundles of benefits" that satisfy their immediate wants. Products/services that producers regard to be

Table 4.3. Industrial versus consumer products/services

Critical factors	Industrial products/ services	Consumer products/ services
1. Number of buyers	Few	Many
2. Target end users	Employers	Individual
3. Nature of products/services	Tailor-made, technical	Commodity, nontechnical
4. Buyer sophistication	High	Low
5. Buying factors	Technical, quality, price, delivery, service	Price, convenience, packaging, brand
6. Consumption	OEM parts for reselling, own consumption	Direct consumption
7. Producer end-user contact	Low	High
8. Time lag between demand and supply	Large	Small
9. Segmentation techniques	SIC, size, geography, end user, decision level	Demographics, lifestyle, geography, ethnic, religious, neighborhood, behavior
10. Classification of goods	Raw materials, fabricated parts, capital goods, accessory equipment, MRO supplies	Convenience (household supplies, foods), shopping (cameras, refrigerators), specialty (brand-name items)

MRO, maintenance, repair, and operations; SIC, standard industrial classification.

different because of physical embodiment (e.g., input materials), production process, or functional characteristics may in fact be equivalent in how consumers perceive them, provided that the same or similar benefits (substitute services) are recognized by the customers. Table 4.4 contains illustrative examples of these different perceptions.

Table 4.4. Service perceptions

Services	Vendor's perceptions	Consumer's perceptions
Major surgeries	Sequences of diagnostic tests	Hope of recovering while enduring pains and suffering
	Surgical procedure	
	Medications, emergency steps	
Financial advisement	Models, diversification strategy	Chances of preserving capital and making money
	Economic projections	
	Projected risks in global economy	

Companies must define competition based on the way customers perceive their products/services. Note that services that appear to be physically different to marketers may appear to be the same to users.

Product/service strategy must also be established with respect to competition. A company may decide to market premium products/services, characterized by having features that are novel or superior to those offered by the competition. Such outstanding product/service features may be possible because of the company's innovative capabilities, technological superiority, and other core competencies. Companies with such "high-road" brands (see Section 4.5.1[E]) tend to enjoy and sustain high profitability. Other companies may elect to make commodity-type (value) products/services with commonly available features so that they compete head-on against their competitors on the basis of price, service, distribution, and customer relations management. They pursue the option of "low-road" products/services. Positioning is the step that addresses such competitive issues related to products/services.

A. Product (Service) Composition

When customers buy and use a product/service, they realize specific benefits from the core offering, as well as a number of supplemental values that support the core. Lovelock and Wirtz[25] introduce the "Flower of Service" model to illustrate the composition of such a service; see Figure 4.9. While this model was created for hotal servcies, it applies to many other

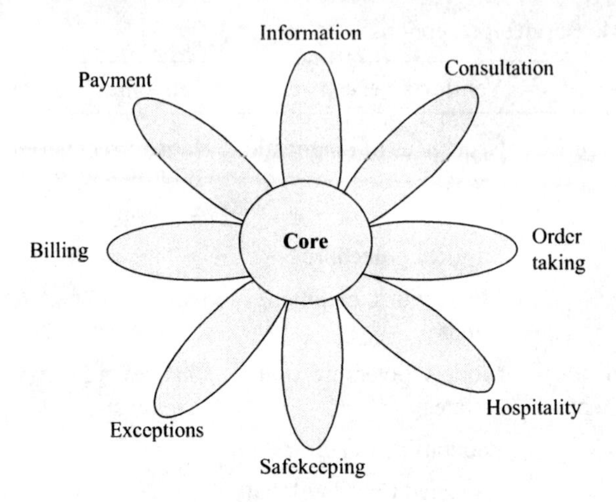

Figure 4.9. Flower of service model.

products/services as well, even though not all supplemental services listed are of equal importance to others. In the clockwise direction, the first four supplemental services are normally expected from any core, whereas the last four may vary depending on the specific core prodcuts/services at hand.

1. **Information:** must be timely and accurate in specifying service hours, locations, price, usage instructions, conditions of sales, warnings, directions to service sites, receipts, tickets, documentation, tracking of package delivery by FedEx front-line employees, technological enablers—website, video kiosks, brochures, etc.
2. **Order-taking**: make the process easy and timely for customers to receive order confirmation, assure the order being accepted quickly, and get receipts.
3. **Billing:** must be timely, accurate, itemized, and clear in order to promote fast customer payment. Set up a system to encourage self-billing and promote fast checkout.
4. **Payment:** make it easy and convenient for customers, such as allowing the use of credit cards and electronic payment options.
5. **Consultation:** must be customized to the situation at hand for dispensing advice related to the use of product/service, counseling, offering free-of-charge tutoring/training or business advisement, as a way to promote sales.
6. **Hospitality:** must be friendly in interacting with customers. Airlines offer departure and arrival lounges to enhance customer convenience. Physical appearance of such facilities is important.

7. **Safekeeping:** This refers to safes in hotel rooms, lighting of parking spaces for cars on site, coat rooms, child care and pet care facilities, and floor management to assure customer safety.
8. **Exceptions:** These include special requests, problem solving, complaint handling, and restitution. Oftentimes, supplemental services are used to define differentiable grades of service, such as premium, standard, economy, and so on. For the service to be successful in the marketplace, the attitudes and approach of customer-facing personnel must be friendly with well-planned customer engagement strategies.

B. Product/Service Positioning

An important question that companies should answer is, which service attributes should be included? A *perceptual map* is a useful tool to position the company's products/services in relation to existing competition in the marketplace. It enables companies to select the correct set of product/service attributes to maximize their marketing advantages. Useful steps in positioning new entries or repositioning existing products/servicess may also include articulating customer preferences and emphasizing gaps in attributes.

Figure 4.10 is an example of a perceptual map for four-year B.S. degree programs in Engineering considering the service attributes of value and cost. Value is defined as knowledge and skills gained plus the intangible benefits (e.g., social networking, alumni organization activities, school

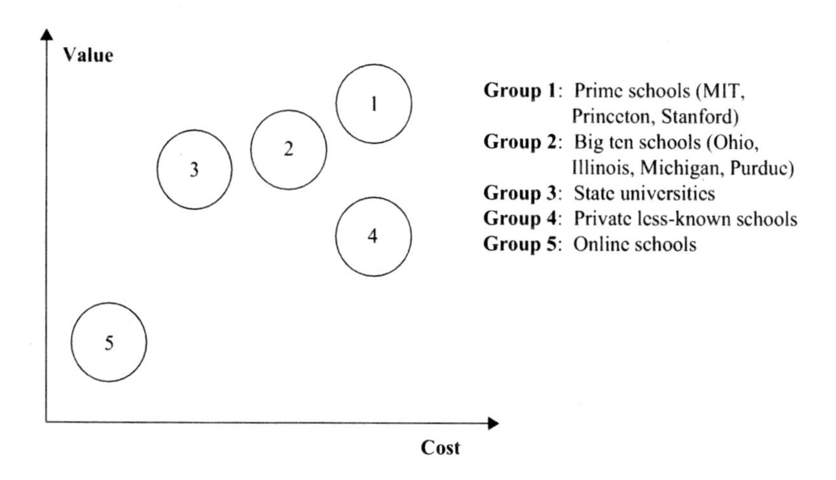

Figure 4.10. Perceptual map.

reputations) related to the institutions involved. Cost include tuitions (out of state), living expense, and others required to complete the four-year programs. Only the relative magnitudes of the attributes are emphasized in such a map. However, the map helps to identify which degree programs are in direct competition and which are not. It is also possible to link customer segments to these pairs of service attributes, thus enabling the product/servcie providers to refine their advertising strategies for these customer segments.

Services with more than two important attributes are readily mapped into an n-dimensional perceptual map. A service (e.g., S1) is designated by a single point having the coordinates F_1, F_2, F_3, through F_n, with each representing an independent service attribute. This representation is complete if the elements of the attributes set (F_1, F_2,..., F_n) are mutually exclusive and collectively exhaustive. For example, for automobiles, these attributes may include price, styling, fuel economy, driving comfort, safety, brand prestige, power, and longer-term dependability (e.g., number of problems per 100 three-year-old vehicles). The spacing between two neighboring points (each identifying a service) as depicted in this n-dimensional map is equal to the square root of the sum of the individual attribute differences, squared. Presented in Table 4.5 is an example of the description of services with six distinguishable attributes.

C. Product/Service Life Cycles

Every product/service goes through a number of important stages throughout its useful life.[26] These stages include:

Table 4.5. Inputs to six-dimensional perceptual map

Service	F_1	F_2	F_3	F_4	F_5	F_6
S1						
S2						
S3						
S4						
Your Service						
S5						
S6						
S7						

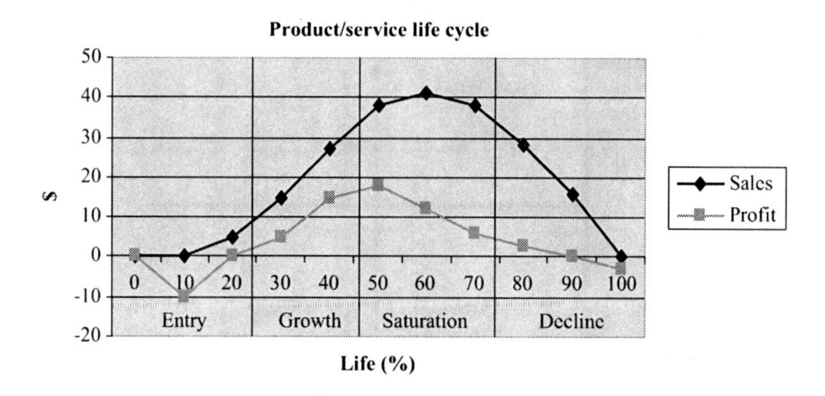

Figure 4.11. Product/service life cycle.

1. The *entry stage*: product/service testing, market development, and advertising
2. The *growth stage*: product/service promotion, market acceptance, and profit growth
3. The *stagnation stage*: price competition, substitution, and new technologies
4. The *decline stage*: cash-cow strategy with no more investment.

Companies need to understand which stage a given product/service is currently in (see Figure 4.11). From the standpoint of the product/service life cycle, an important product/service development strategy is to sequence the introduction of new products/services so that a high average level of profitability can be maintained for the company over time.

Engineering managers are particularly qualified to constantly come up with innovative products/services so that their employers may introduce these offerings at the right intervals.

D. Product/Service Portfolio

Another strategy issue is related to the types of products/services concurrently being marketed by the conpany. With the exception of a few, most companies market a group of products/services at the same time, referred to as a *portfolio.*[27] Products/services in a portfolio are usually not "created equal." From the company's standpoints of pursuing profitability and market-share position, some are more valuable than others. *Boston Consulting Group* (BCG) of Boston, Massachusetts, developed a portfolio matrix based on the measures of growth rate and market share

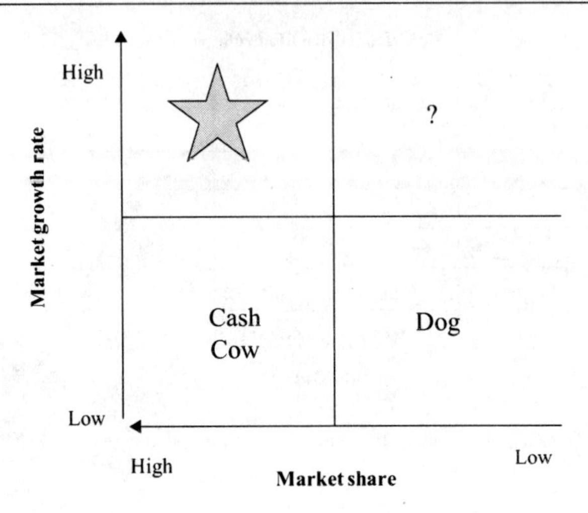

Figure 4.12. Product (service) portfolio.

(see Figure 4.12). According to this classification scheme, products/services are regarded as *stars* if they enjoy high growth rate and high market share. They are *question marks* if achieving have high growth rate, but low market share. *Cash cows* are those products/services with low growth rate and high market share. Products/services are designated as *dogs* if both growth rate and market share are low.

Figure 4.12 indicates that companies need to differentiate the products/services they market by strategically emphasizing some and deemphasizing others, according to the responses from the marketplace. For example, a useful corporate strategy to manage a product/service portfolio is to milk the *cows* to provide capital for building *question marks* into *stars* that will eventually become *cash cows*. Divest the *dogs*.

E. Producs/Services and Brands

Numerous high-tech companies operate in a "service-centric" business model, in that they market products/services on price and performance. Recent market studies show that the success of technology-based products/services in the marketplace is not purely dependent on the price–performance ratio, but also on the trust, reliability, and promised values the customers perceive in a given brand.

According to Dev,[28] brand is "a distinct identity that differentiates a relevant, enduring and credible promise of value associated with a product, service, or organization that indicates the source of that promise."

Table 4.6. Promises of value

Corporate brands	Promises of value
IBM	Superior service and support
Apple	Simple, innovative, and easy to use
Lucent	Newest technologies
Gateway	Friendly service

The brand of a company represents more than a name. It stands for all of the images and experience (e.g., products, services, interactions, management quality, and customer relations) that customers associate with the organization. It is a link forged between the company and the customers. It is a bridge for the company to strengthen relationships with customers.[29]

A promise of value is an expectation of the customer that the company is committed to deliver. Examples of such promises of value from several companies are listed in Table 4.6. They must be relevant to the enduring needs of the target customers and made credible by the persistent commitments of the company. To be competitive, the promise of value must be distinguishable from those offered by other brands.

Research by Ward et al.[30] indicates that customers consider questions at five levels when purchasing both high-tech and consumer types of products/services. These questions may be grouped into a brand pyramid, as illustrated in Figure 4.13.

Technology-oriented buyers are typically focused on questions at levels 1 and 2. However, higher-level business managers who make purchase decisions are also known to address questions at levels 3 to 5. These decision makers are interested in what the product/service will do for them, not just how it works. Consequently, to project a trustworthy and reliable image, to build strong relationships with customers, and to enhance brand loyalty and customer retention, companies pay attention to questions at all five levels. This is the emphasis of brand management.

Brand is a major asset that must be properly managed and constantly strengthened. Useful inputs for brand management are typically secured from customer feedback. Once the market is properly segmented, the promise of values is specifically designed to address the needs of the target segments involved. Actions are then taken to deliver the stated promise of values, and results are constantly collected to monitor progress.

Brand is evaluated on how well it is doing with respect to a number of metrics: (a) delivering according to the customer's desire; (b) relevance

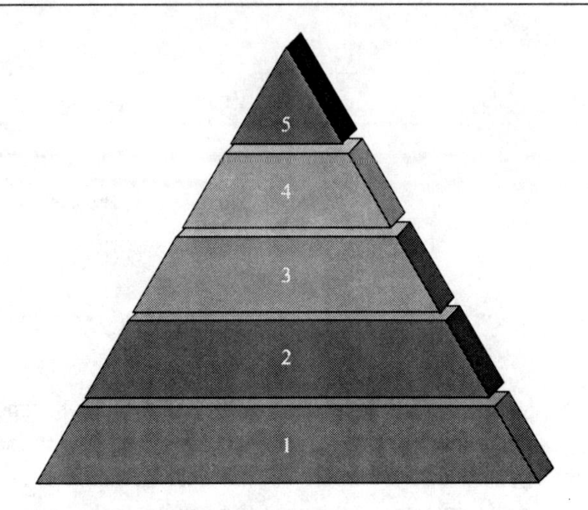

Figure 4.13. Brand pyramid.

Notes: Level 5: What is the personality of the brand (aggressiveness, conservatism, etc.)?
Level 4: Does the value offered by the product/service reflect that favored by the customers (family values, achievement)?
Level 3: What are the psychological or emotional benefits of using the product/service in question? How will the customers feel when they experience its technical benefits?
Level 2: What are the technical benefits to customers (solutions to problems, cost-saving benefits, and speed of production)?
Level 1: How does the product/service work (characteristics, technological features, and functional performance)?

to the customers; (c) value to the customers; (d) positioning; (e) service portfolio management; and (f) integration of marketing efforts, management, support, and monitoring.

In the past, brand management has been focused primarily on points of difference, such as how a given brand differs from the other competing brands within the same category. *Maytag* is known to emphasize "dependability." *Tide*® focuses on "whitening power." *BMW* stresses "superior handling." Recently, Keller et al.[31] suggested that attention should be paid to points of parity and the applicable frame of reference, in addition to the points of difference, when marketing a given brand. Emphasizing the frame of reference is intended to help customers recognize the brand category comprising all of the competing brands. Focusing on the points of parity will ensure that customers recognize a given brand as a member of the identified brand category.

Brand may be classified with respect to the two dimensions: brand category and relative market share. The brand category is defined as

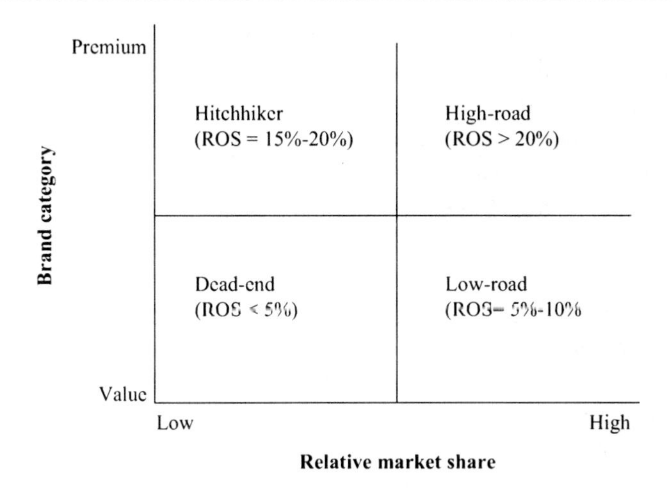

Figure 4.14. Brand classes.

premium if the category is dominated by premium brands—those with high values to customers. Examples of premium brands include *BMW, Mercedes Benz, Jaguar,* and other luxury and specialty cars, each of which has unique , high-value attributes. The brand category is defined as *value* if it is dominated by value brands—those with basic, minimum, low-end attributes. Examples of value brands include *Chevy, Saturn*, and other compact and four-door family cars. *Gillette* markets its Mach-3 Turbo shaving system as a premium brand, whereas the cheap disposable razors from its own company as well as its competition are the value brands. The relative market share refers to the percentage of market share a given brand is able to attain.

In Figure 4.14, brands are grouped into four classes: high-road, low-road, hitchhiker, and dead-end brands.[32] ROS represents return on sales, which is defined as net income divided by the sales revenue (see Section 3.4.1).

High-road brands are those with services that offer premium features, options, qualities, and functions to command high selling prices while attaining a leadership position in the market share. Examples of such high-road brands are *Apple* iPhone 5S, *Microsoft* Surface Pro 2, and *McKinsey & Company*. These brands enjoy excellent profitability that may be sustained for long periods. The key success factors for high-road brands are technological innovation (e.g., constantly adding novel product/service features and values), time to market, flexible manufacturing, and advertising (see Figure 14.3).

Low-road brands are those that offer value brands while enjoying a high market-share position. Because of marketplace competition and a lack of distinguishable service features, these brands can be successfully managed by emphasizing cost reduction, production efficiency, service simplification, and distribution effectiveness.

Hitchhiker brands are those with premium service values and low market share. For these brands to become high-road brands, management must emphasize cost reduction, flexible manufacturing, and product/service innovation.

Dead-end brands are value brands with low market share. These brands attain only marginal profitability. There are several strategies to grow the profitability of these dead-end brands: (a) reduce the price to penetrate the market and thus move these brands to the low-road category; (b) increase the scale of economies by applying the "string-of-pearls" strategy: producing and marketing several services together to cut costs; and finally (c) introduce a superior, premium service to "trump" this brand into the hitchhiker category. Failing all of these attempts, dead-end brands should be discontinued. Table 4.7 summarizes the strategies that deal with these four classes of brands.

The preceding discussion on product/service brands should assist engineering managers in understanding the significant value added by brands to the success of the company's marketing program.[33] Such an understanding should make it easier for them to channel their support efforts to actively enhance the company's brand strategy.

Berry and Seltman[34] presented a useful model to build brand equity. This model is depicted in Figure 4.15. "Organization's presented brand" is the desired brand image advocated by the product/service company. The brand image is promoted through advertising, brand name, logo, websites, employee uniforms and facilities' design. This strategy creates "brand awareness" among target customers. Brand awareness is the customer's ability to recognize and recall a brand. "Brand meaning" is the customer's dominant perception of the brand. Perception is built on the personal experience customers have in dealing with the product/service organization. Customer experience will also influence what other people say and write (e.g., word of mouth, blogs, message boards, etc.) about the brand—these "external brand communications" have a secondary influence on both brand awareness and brand meaning, as depicted by the dotted arrows in Figure 4.15. Together, brand awareness and brand meaning form the brand equity, which is the degree of marketing advantage or disadvantage the organization has over its competition in the marketplace.

Table 4.7. Strategic options for brands*

Brands	Strategic options
High road	Apply R&D to constantly innovate to make products/services premium—adding new product/service features and changing forms and functions
	Expand product/service lines (product/service proliferation)
	Initiate media campaign
	Capital investment
	Decrease time to market
	Flexible manufacturing
	Direct store delivery to preoccupy shelf space
Low road	Pursue cost reduction aggressively
	Lessen product/service proliferation (stock keeping units [SKUs]) by simplifying types and designs
	Consolidate production facilities to improve efficiency and cut wastes
	Use realized cost savings to slash price
	Consider ways to add premium products/services (advancing to high road)
Hitchhiker	Apply R&D to constantly innovate to make products/services premium—adding new product/service features and changing forms and functions
	Cost reduction
	Reduce time to market
	Flexible manufacturing
	Initiate media campaign
	Consider capital investment
Dead end	Cut price (advancing to low road)
	Outsource in areas with economies of scale
	Apply the "string-of-pearls" strategy to enhance scale
	"Trump" the category by introducing a superior, premium product/service that resets consumer's expectation (advancing to hitchhiker)
	Do not spend on marketing
	Make no capital investment

**Source*: Vishwanath and Mark (1997).

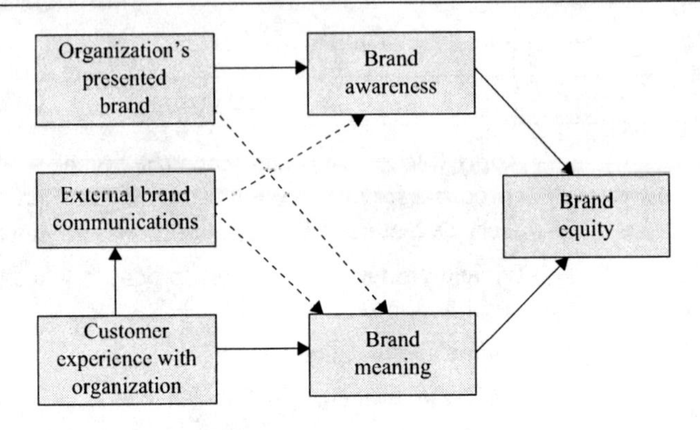

Figure 4.15. Model of building brand equity.

Based on this model, building a strong brand equity requires (a) dedicated effort in brand presentation (e.g., advertising, logo, websites, messages, physical appearance of facilities); (b) commitment to effect a satisfactory customer experience (e.g., employee training, employee compensation, employee attitudes, efficient support processes, pleasant working environment, etc.); and (c) a customer feedback system to allow efficient mitigation of any negative external communications (e.g., unfavorable publicity, ethics, corporate citizenship). In services, the customer's experience is typically generated by interacting with customer-facing agents. The brand name is associated with the entire service organization, including all service people and products.

The *Mayo Clinic* is a world-renowned hospital headquartered in Rochester, Minnesota. Its brand name is built on three specific strategies: (a) unique skills and clinical outcome, (b) a team medicine model to solve patients' problems collaboratively, and (c) actual experience exceeds expectations. The Mayo Clinic, by emphasizing the value of services it offers and creating a system capable of meeting and surpassing expectations, was able to turn its patients into marketers for the clinic. Because the Mayo Clinic treats its patients well, patients would talk to 40 others about their good experience at the clinic. The marketing promotion effected by its own patients has been possible only because the outcomes the Mayo Clinic projected exceeded patient's expectations. Just meeting expectation is not enough to generate a word-of-mouth campaign.[35]

Brand equity is the sum of a customer's assessment of the brand's intangible qualities, positive or negative. According to Rust et al.,[36] emphasizing only the building of brand equity misses valuable opportunities, often to the detriment of the company. Instead of focusing on brand equity, companies should focus on customer equity. The logic is simple. A customer may change his or her perception about a specific brand. However, if these customers can be encouraged to attach their loyalty to another brand within the company, their relationship with the company will continue. Customer life time value is the net profit derived from a customer during the time when the customer has a positive relationship with the company.

Brand perception can be influenced by geography. For example, South Americans may have a vastly different perception about an American brand than people in the United States. The following are recommended strategies[37] to pursue customer-centric branding strategies:

1. Focus on customer relationships ahead of branding. Manage key accounts of profitable and important customers and apply appropriate branding strategies to these customer groups or segments.
2. Build brand around customer segments, not the other way around.
3. Make the brand as narrow as possible. In a customer-centered approach, the brand should be able to satisfy small customer segments while remaining economically viable.
4. If the customers are similar, then different brands may be advertised to them. An Example is Disney, which offers movies, hotels, and amusement parks all to the same customers (i.e., the young and young at heart, all who want to be entertained).
5. Be ready to hand off customers to other brands in the same company. For example, encourage customers of Fairfield Inn to trade up to the Marriot Hotel brand (Marriot owns the Fairfield Inn brand).
6. Do not take heroic measures. If a brand is no longer viable, do not spend time and resources attempting to save it. Let it go.
7. Change how brand equity is measured.

The long-term goal is to create and cultivate profitable, long-term relationships with customers. Brand management is only one tool of many that can be used toward accomplishing this goal.

Example 4.3

Forecasting future market conditions and technologies is a difficult, but necessary task for companies striving to sustain business success. Looking out for emerging technologies, which could be applied to enhance business competitiveness, should be the primary role of engineering and technology managers.

What might be a good strategy for engineering and technology managers to become sensitized to forecasting technology and scanning emerging technologies so that they fulfill their important role of serving as technology "gatekeepers" to their employers?

Answer 4.3

Different engineering and technology managers will have different preferences in fulfilling this important role of forecasting. One possibility is to adopt the following logical sequence of steps:

1. Compose a "wish list" of technologies that would make the company's current products/services cheaper, faster, and better. Define desirable new product/service features based on customer inputs and the technologies required for their development. Define new product/service concepts and the requisite technologies that might be compatible with the current product/service lines marketed by the company.
2. Understand some of the emerging technologies noted in the literature. Network actively with professionals in industries, universities and independent research organizations to access valuable information sources.
3. Determine the useful technologies that might be available during the next 5 to 10 years to support the current products, product enrichments, or new product/service concepts.
4. Assess the development activities associated with these useful technologies in universities, start-ups, technology incubator firms, contract research companies, or other organizations, both domestic and global, to gauge their quality and readiness for commercialization.
5. Make specific recommendations in a timely manner to secure the supply of such new technologies, by way of acquisition, joint development, and partnership, for enhancing the commercial success of the company's products/services.

Example 4.4

There is a strategic approach called "Second Brand strategy." Explain, what is unique about this strategy? Under what conditions would this strategy be best applied and what is the main purpose it is intended to achieve?

Answer 4.4

The "Second Brand Strategy" refers to the creation of a new brand, which is to be distinguished from the primary brands of the company, in order to market similar products at much lower prices. Companies pursue this strategy to avoid or minimize the cannibalization effect on the primary brands and to counter the aggressive selling efforts of new competitors who enter the marketplace with low-price alternatives.

This strategy would work well for companies that have financial staying power, are presently the market leaders interested in protecting their market share positions, and want to pursue this strategy as a short-term solution to confront the market penetration efforts of new competitors and drive them away, while preserving the loyalty of their current customers. As soon as the competitors become disenchanted and disappear from the scene, the company would stop the "Second Brand Strategy" immediately.

Example 4.5

XYZ Company wishes to enter a new market arena on the basis of its strength in core technologies and financial staying power. However, the market arena in question is currently dominated by a major competitor with 80 percent of the market share, and a number of smaller competitors, each being focused on small niche segments. How should XYZ Company enter this market?

Answer 4.5

One strategy is to (a) acquire a couple of small competitors to establish a foothold in the target market arena, (b) invest to manufacture new products built upon one's own core technologies, (c) consider importing suitable parts to cut product costs and speed up the time to market, and (d) sell the products at slightly lower prices to gradually take the market share away from the dominant player.

F. Engineering Contributions to Product/Service and Brand Strategy

The product/service is a key element in the marketing mix. Engineers and engineering managers have major opportunities to add value by

(a) understanding the customers' perceptions of products/services; (b) designing products/services with features that are wanted by customers; (c) helping to position the company's services strategically to derive marketing advantage; (d) practicing innovations in the design, development, production, reliability, serviceability, and maintenance of products/services to differentiate them from others; (e) sustaining the company's long-term profitability by creating a constant flow of new products/services for introduction on a timely basis; (f) assisting in managing companies' product/service portfolios by adding premium features to some and reducing costs to others; and (g) ensuring commercial success of the high-road and hitchhiker brands in the marketplace.

In the "knowledge economy" of the 21st century, time to market is an increasingly important competitive factor. Once the desirable set of product/service attributes is known from market research, those companies that bring the suitable products/services to the market first will enjoy preemptive selling advantages and will recover the product/service development costs faster than others.

Engineering managers should also be well prepared to contribute in shortening the services' time to market by utilizing advanced technologies (e.g., 3-D printing, nano-technologies, etc.) to create modular design, eliminate prototyping, whittle away design changes, foster parts interchangeability, ensure quality control, and introduce other innovations.

4.5.2 PRICING STRATEGY

Price is a very important product/service attribute.[38] Companies pay a great deal of attention to the setting of prices. Setting the price too high will discourage consumers from buying, whereas setting it too low will not assure profitability for the company. In general, companies have a number of options to define its pricing strategy, as described below.

A. Skimming and Penetration Strategies

Companies applying the *skimming strategy* would set the product/service price at the premium levels initially and then reduce it in time to reach additional customers. In other words, they "skim the cream" first. An example is the marketing of a new book with hardcover copies selling at a high price (e.g., \$29.95) followed by the paperback version at a low price (\$4.95).

New technology services are also typically sold at high prices initially in the absence of competition. As competitors enter the market with products/services of similar features and quality, product/service prices are then lowered accordingly.

In contrast, companies pursuing the *penetration strategy* would set product/service prices low to penetrate a new market and to forcefully acquire a large market share. A high market-share position sets forth a barrier of entry for other potential competitors. Typically, companies use a penetration pricing strategy to enter an existing, but highly competitive market. An example is the marketing of Japanese motorcycles in the United States.

B. Factors Affecting Price

In setting product/service prices, besides using the skimming and penetration strategies, companies broadly consider a number of other factors: financial aspects, service characteristics, marketplace competition, distribution and production capabilities, price–quality relationship, and the relative position of power. These factors will be discussed next.

1. **Financial aspects:** The more solid the company's financial position is, the more capable it is to initially set the product/service price low. Companies strong in finance stay afloat for a long period of time even with low profitability. Companies that desire high, short-term profitability tend to set the product/service prices high.
2. **Product/service characteristics:** The product/service price may be set in accordance to its value and importance to users, as well as the income levels of its target customers. Usually, a new product/service in its early life cycle would sell at a high price to allow the company to benefit from the product/service's novelty.
3. **Market place competiton:** Companies set product/service prices in reverse proportion to the level of competition in the marketplace. The level of competition refers to the number of direct competitors, the number of indirect competitors marketing substitution products/services that offer similar value to customers, and the counterstrategies (e.g., speed and intensity) that these competitors may implement. Companies tend to set the product/service price high if the barriers to market entry are high. The barriers to market entry depend on lead time and resources—technical and financial, patents, cost structure, and production experience. In addition,

products/services with inelastic price–demand characteristics tend to carry a high price. A product/service has inelastic price–demand characteristics if a large price increase induces a small change in the quantity of the product/service demanded in the marketplace.

4. **Distribution and production capabilities:** Service availability to consumers depends on the company's product/service distribution capabilities. With strong distribution channels in place, companies may set the product/service price high, as quickly making products/services available to consumers represents a competitive strength.

   Sales volume impacts the company's production experience. Companies with extensive production experience are known to produce products/services at a low unit cost. A lower product/service unit cost enables these companies to set a lower price to gain market share. *The Boston Consultant Group* studied manufacturing operations and discovered that there is a correlation between production volume and product/service unit cost. For every doubling of the production volume, the unit cost is whittled down by about 15 percent—or the 85 percent experience curve.

   Companies with a faster time-to-market strategy are able to accumulate production volume more quickly, attain a lower product/service unit cost sooner, and sustain company profitability for longer periods of time.

5. **Price–quality relationship:** One important consideration in setting the product/service price is the perceived cost–quality relationship by customers. There is substantial evidence in business literature to indicate that customers tend to believe that "low-priced items cannot be good." Price is perceived to be an important indicator of quality.

   Therefore, product/service prices should not be set too low. There is a price threshold below which customers may raise questions regarding the product/service quality, as indicated in Figure 4.16. The demand curve "Quantity A" illustrates a normal price–demand relationship in the absence of a price threshold, whereas the demand curve "Quantity *T*" contains a price threshold at about $30 per unit, below which the demand for the products/services in question starts to drop off as the perception of poor quality related to low price sets in.

6. **Relative position of power:** Products/services for consumers are typically marketed by a few major companies to a very large number of customers. On the other hand, for industrial products/services with high technological contents, the number of both producers and

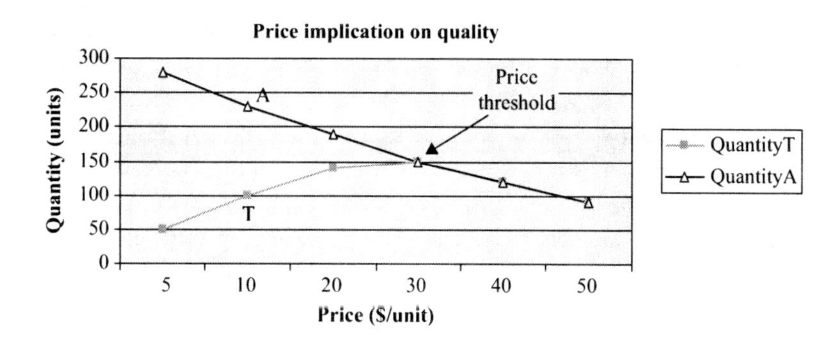

Figure 4.16. Price–quality relationships.

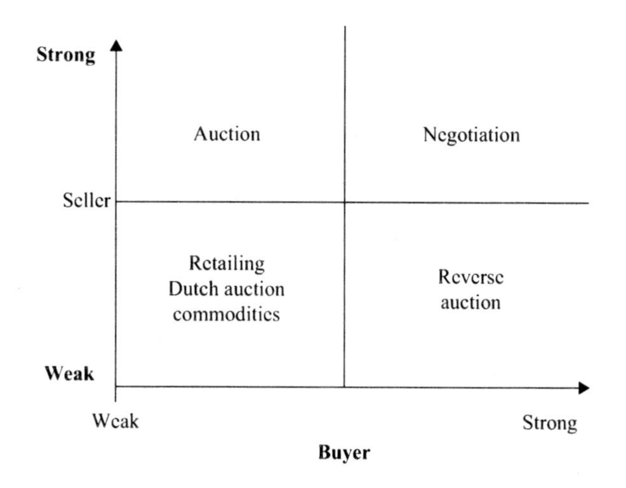

Figure 4.17. Processes of setting product/service price.

customers may be limited. The greater the number of sellers there are available for a given product/service, the weaker will be each seller's position in the marketplace. Similarly, the more buyers exist for a specific product/service, the weaker will be the buyers' relative market position.

Less competition makes either sellers or buyers more powerful. The relative position power between buyers (customers) and sellers (producers) has an impact on product/service pricing, as illustrated in Figure 4.17. The final price offered by the sellers and accepted by the buyers is usually arrived at by a suitable negotiation or auction process.

If both buyers and sellers are strong—for example, when the US government (customer) procures fighter airplanes from the defense industry (producer)—a final price is typically reached by a *negotiation* made up of a series of offers and counteroffers. A typical pricing arrangement may be cost plus a fixed percentage of gross margins.

When the sellers are strong (e.g., selling an original master painting, a porcelain vase from the Ming Dynasty, or some other type of unique physical asset) and the buyers are weak, sellers tend to take advantage of their dominant supplier position by employing an auction. An *auction* is a bidding process by which buyers are forced to compete against each other by committing themselves to consecutively higher prices, with the final price being set by the highest winning bid.

If buyers are in a strong position (e.g., due to large transaction volumes), they force weak sellers to compete against each other in a reverse auction. A *reverse auction* requires the prequalified sellers to submit increasingly lower bid prices within a fixed period of time. The lowest bid defines the final price and the ultimate winner of the sales contract.[39] Some large companies employ such pricing tactics to purchase high-volume supply items such as computers; paper and pencils; tires; batteries; and maintenance, repair, and operations (MRO) goods.

Finally, when both sellers and buyers are weak, products/services are usually not differentiable, and the prices are highly depressed and fixed. Examples include various commodity products/services sold in retail stores. Some sellers (e.g., Land's End) may activate a Dutch auction to compete. In a *Dutch auction*, sellers slash the product/service prices consecutively by certain percentage at a regular time interval (e.g., every week) until the products/services are sold or taken off the market. In this case, buyers compete against other "sight-unseen" buyers to seize the lowest possible selling prices.

The Internet has made many of these pricing processes much more practical and efficient to implement.[40] Because of its ability to allow sellers and buyers to rapidly reach other buyers and sellers, respectively, the Internet tends to weaken the relative power positions of both the sellers and the buyers, causing services to become increasingly commoditized, thus depressing product/service prices and intensifying competition.

Table 4.8 enumerates a number of other factors that have an impact on setting the product/service price.

Table 4.8. Factors affecting service price

Factor	Skimming	Penetration
Demand	Inelastic	Elastic
	Users know little about service	Familiar service
	Market segments with different price elasticity	Absence of high-price segment
Competition	Few competitors	Keep out competition
	Attracts competition	Market entry easy
	Market entry difficult	
Objective	Risk aversion	Risk taking
	Go for profits	Go for market share
Service	Establish high-volume image	Image less important
	Service needs to be tested	Few technical service problems
	Short service life cycle	Long service life cycle
Price	Easy to go down later	Tough to increase later
	More room to maneuver	Little room to maneuver
Distribution and promotion	No previous experience	Existing distribution system and promotion program
	Need gross margin to finance its development	
Financing	Low investment	High investment
	Faster profits	Slower profits
Production	Little economy of scale	High economy of scale
	Little knowledge of costs	Good knowledge of costs

C. Pricing Methods

In setting product/service prices, companies broadly consider a number of factors and methods. Several of these methods are briefly discussed next.

1. **Cost oriented:** Some companies set prices by adding a well-defined markup percentage to the product/service cost. This is to ensure that

all products/services sold will generate an equal amount of contribution margin to the company's profitability:

$$\text{Price} = \text{cost} + \text{markup (e.g., 35 percent of cost)} \quad (4.1)$$

Cost-plus contracts are often used for industrial products/services related to R&D, military procurements, unique machine tools, and others. Small sellers use cost-plus pricing to ensure a fair return while minimizing cost factor risks. Larger buyers favor this type of pricing so that they can push for vendor cost reduction via experience. Larger buyers may optionally offer to help absorb the cost risks related to inflation.

Often, sellers and buyers enter a target-incentive contract, which prescribes that, if actual costs are lower than the contract costs, sellers and buyers split the savings at a specific ratio. On the other hand, if the contract costs are exceeded, then both parties pay a fixed percentage of the excess; the buyers pay no more than a predetermined ceiling price.

2. **Profit oriented:** Other companies prefer to require that all products/services contribute a fixed amount of profit. This pricing method ensures that sellers realize a predetermined return-on-investment (ROI) goal:

$$\text{Price} = \text{cost} + \text{profits (e.g., in terms of ROI)} \quad (4.2)$$

3. **Market oriented:** Some companies set prices of certain industrial products/services, such as those requiring customization, to what the buyers are willing to pay. For example, the companies strive to negotiate for the highest price possible and take advantage of the fact that product/service and pricing information may not be easily accessible. The continued advancement of Web-based communication tools tends to make information just one click away, rendering this type of pricing method no longer practical in today's marketplace.

   Companies may also price the products/services slightly below *the next-best alternative* products/services available to the customer. The companies that have exhaustively studied the next-best alternative products/services available to their customers garner advantages in price negotiations.

   Competitive bidding is used often by governments and powerful buyers. Usually three bids are needed for procurements exceeding a specific dollar value. Sealed bids are opened at a predetermined date, and the lowest bidder is typically the winner. Some industrial companies may engage in *negotiated bidding*, wherein they continue negotiating with the lowest one or two bidders for additional price concessions after the bidding process (e.g., a reverse auction) has been completed.

Value-added pricing—Companies with extensive application know-how related to their industrial products/services may set prices in proportion to the products/services' expected value to the customer. The product/service's value to the customer depends on the realizable improvement in quality, productivity enhancement, cost reduction, profitability increase, and other such benefits attributable to the use of the product/service in question. Producers set the prices high if there is a large value added to the customer by the use of their products/services.

4. **Competition oriented:** A common pricing method is to set prices at the same level as those charged by the competitor. Doing so induces a head-on competition in the marketplace. In oligopolistic markets (typically dominated by one or two major producers or sellers and participated by several other smaller followers), the market leader sets the price.

   One well-known example of a competition-oriented pricing practice is target pricing. *Target pricing* was initiated and applied by many Japanese companies. Some American companies have now started to successfully apply this method. The process of target pricing (see Figure 4.18) is as follows:

   (a) Determine the market prices of products/services that are similar or equivalent to the new product/service planned for introduction by the company. Find all product/service attributes customers may desire. This is usually accomplished by a multifunctional team composed of representatives of such disciplines as design, engineering, production, service, reliability, and marketing. Select a product/service price (e.g., at 80 percent of the market price) that makes the company's new product/service competitive in the marketplace. This is then the target product/service price.

   (b) Define a gross margin that the company must have in order to remain in business.

   (c) Calculate the maximum cost of goods sold (CGS) by subtracting the gross margin from the target product/service price. This is the target product/service cost, which must not be exceeded.

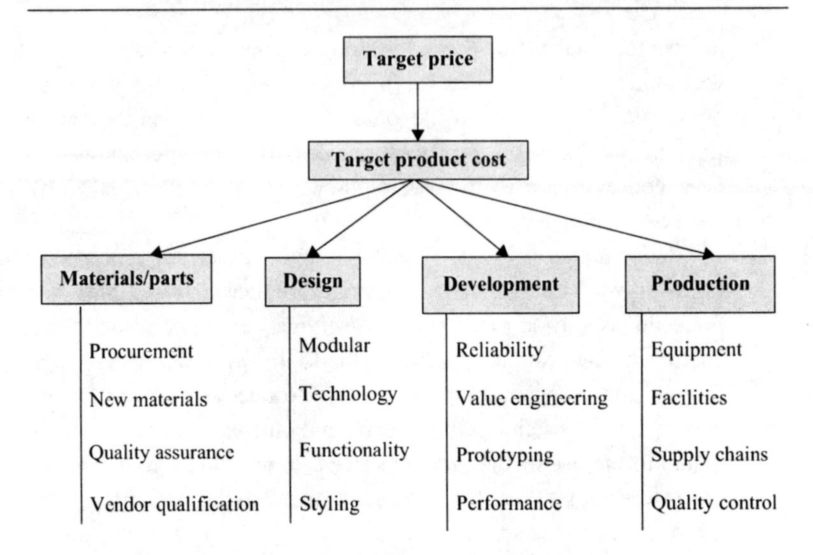

Figure 4.18. Target pricing.

(d) Conduct a detailed cost analysis to determine the costs of all materials, parts, subassemblies, engineering, and other activities required to produce the new product/service. Usually, the sum of these individual costs will exceed the target product/service cost previously defined.

Apply innovations in product/service design, manufacturing, procurement, outsourcing, and other cost-reduction techniques to bring the CGS down to or below the target product/service cost level.

(e) Initiate the development process for the new product/service only if the target product/service cost goal can be met.

The target pricing method ensures that the company's new product/service can be sold in the marketplace at the predetermined competitive price, with features desired by consumers, to generate a well-defined profitability for the company. This method systematically evaluates low-risk, high-return investment opportunities because it forces the company to invest only when the commercial success of the service is more or less assured. Furthermore, it also focuses the company's product/service innovations on finding ways to meet specific and well-defined target cost goals. It avoids the potential of wasting its precious intellectual talents and financial resoruces in chasing ideas with no practical value.

Numerous companies use the pricing methods just discussed. Product/service prices are usually set by the marketing department in consultation with business managers. Engineering managers are advised to become aware of these methods, but to defer pricing decisions and related discussions to the marketing department.

D. Pricing and Psychology of Consumption

Recent studies indicate that buyers are more likely to consume a product/service when they are aware of its cost. The more they consume, the more they will buy again and thus become repeat customers. One useful way to induce them to repeatedly consume the product/services is to remind them of the costs committed through the choice of payment methods. This is based on the assumption that the more often the customers are reminded of the payments, the more they feel guilty if they do not fully utilize the products/services they have paid for.

It was pointed out in the literature that time payment will better induce regular consumption of a product/service than lump-sum prepayment (e.g., prepaid season tickets) at the same total value.[41] This is because the time payments remind the buyers of the costs periodically and thus invoke the *sunk-cost effect* on a regular basis. The psychology of the sunk-cost effect is that consumers feel compelled to use products/services they have paid for to avoid the embarrassing feeling that they have wasted their money.

Credit card payments are less effective in inducing consumption than cash payments because cash payments require the buyers to take out currency notes and count them one by one; thus, they experience the "pain" of making payments.

In price-bundling situations, the more clearly the individual prices of products/services are itemized, the better the perceived sunk-cost effect will be. Breaking down large payments into a number of smaller ones, thus clearly highlighting the costs of individual products/services sold in the bundle, can enhance this effect.

Studies of membership rates at commercial wellness and fitness centers support this logic. It has been well documented that those members who pay the membership fees once a year use the facilities only occasionally. These members are the least likely to renew, in comparison with those who pay on a monthly basis. Similar observations are made in sports events in which holders of season tickets show up less frequently than those who buy tickets for specific sets of events.

These examples point out that companies can induce customers to become repeat customers by focusing on ways of encouraging consumption. Only consumption lets customers experience the benefits of the products/services they have purchased. Without such favorable experience, they may not feel that they have good reasons to buy the products/services again in the future. Hence, besides providing a good bundle of value made up of price, product/service features, convenient and efficient order processing and delivery, and quick-response after-sales services, companies should devise ways to stimulate consumption as a strategy to cultivate and retain repeat customers.

Example 4.6

The company has been selling a number of services to customers. It is about to launch a new service with features far superior to any services currently in the market. One option is to price this new service at a large premium above the current price range so that its heavy development expenses can be readily recouped. The other option is to set the price comparable with the existing range in order to retain customer loyalty. What is your pricing advice to the company?

Answer 4.6

Hold a focus group to find out the potential response of customers to the new service's features. Are these features of real value to them? How much more are they willing to pay for these features? Exciting new features from the manufacturer's viewpoints may not be as exciting to customers. Should customers appreciate the new features, then it is advisable to apply the skimming strategy and set a high price for the new service. This is also the principle of value pricing. Furthermore, doing so will avoid "cannibalizing" the existing services of the company.

An efficacious promotional campaign is essential to heighten service awareness. Keep monitoring the response of the market. If the market response is poor, cut down the service price slowly to induce more demand.

4.5.3 PROMOTION (MARKETING COMMUNICATIONS)

Regarding *promotion* and *communication,* companies consider strategies of product/service and brand promotion, options to use a push–pull strategy, selection of advertising media, and the choice of promotional intensity. These considerations ensure that the selected means for communication are compatible with the characteristics of the target market

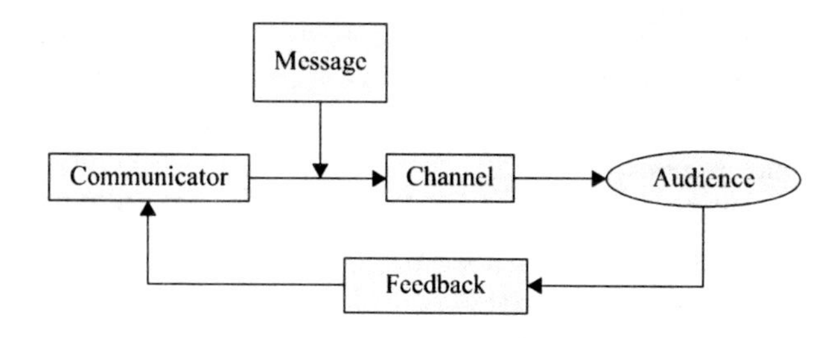

Figure 4.19. Marketing communications

segments. Marketing communication is intended to make the target customers aware of the features and benefits of the company's products/services. Promotion needs to follow a well-planned process (see Figure 4.19) by defining who says what to whom, in what way, through what channel, and with what resulting effects.

A. Communication Process

Companies select communicators who are publicly recognized and have trustworthy images, as these speakers tend to induce public acceptance of their messages. Examples include Scarlett Johansson for SodaStream, Stephen Colbert for pistachios, Ben Kinsley for F-Type Coup of Jaguar, and Arnold Schwarzenegger for Bud Light (see 2014 Super Bowl 30 seconds commercials costing $4 million each).

Messages may be in various forms including slogans. A slogan is a brief phrase used to get the consumer's attention and acts as a mnemonic aid. Successful slogans typically represent a symbolization of product/service features in terms of the customer's wants and needs (such as information, persuasion, and education). Examples include "Ring around the collar," "Where's the Beef?" "You are what you know," "One investor at a time," and Chrysler's "Imported from Detroit."

Channels of communication are specific avenues to foster market communications. In general, there are two types of channels of communications: the marketer controlled and the consumer controlled. The *marketer-controlled* channels include advertisements placed in trade journals, national television programs, distribution of specific service brochures, promotion by technical salespeople, industrial exhibitions, and direct-mail marketing. The *consumer-controlled* channels include interpersonal communications by word of mouth, news reports, and other sources of information perceived to contain no conflicts of interest.

The audience is the target for marketing communications. When selecting communication channels to reach specific consumer segments, the segments' characteristics, media habits, and service knowledge must be taken into account. Segments' characteristics include socioeconomic status, demographics, lifestyle, and psychology. For industrial customers, segment characteristics include big versus small firms, large versus small market shares, and stable versus unstable financial position. Media habits point to sources of information preferred by the customers (e.g., types of magazines and TV programs). Product/service knowledge is the consumers' understanding and appreciation of the values offered by the product/service packages.

Some companies invest a considerable amount of efforts into educating their consumers. A case in point is the known practice of some drug companies of sponsoring large-scale clinical studies conducted by universities and other independent organizations. The purpose of such funded studies is to produce results for publication in technical journals from which consumers gain product/service knowledge in ways preferred by the sponsoring drug companies.

The impact of marketing communications on products/services may be short term or long term. The short-term impact is related to recall, recognition, awareness, and purchase intention with respect to the products/services in question. The long-term impact is reflected in the purchase by customers and brand loyalty with repeat purchase. Several factors influence the effectiveness of marketing communication, such as timing, price, product/service availability, responses by competition, product/service warranty conditions, and support service.

Marketing communication brings about heightened product/service awareness. An improved familiarity with the product/service induces more people to buy the offerings at the current price, thus causing an up-shift of the product/service demand curve (as illustrated in Figure 4.20).

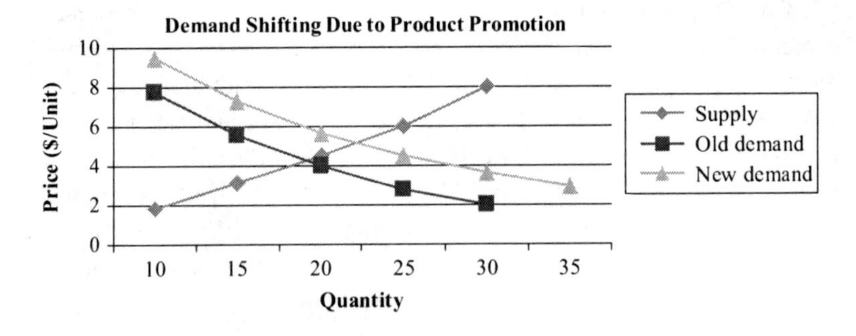

Figure 4.20. Uplift of demand curve.

B. Promotion Strategy

Product/service promotion may be pursued by either causing the consumers to want to pull the products/services from the supply chain or pushing the products/services to the consumers through the supply chains. Many companies practice both strategies.

In a *pull* strategy, the consumers go to retail stores to query about the products/services because they have been informed of their values by advertisements and other promotional efforts of the sellers. In this case, the product or service is presold to the consumers, who practically pull the product/service through its distribution channels (see Figure 4.21).

In exercising a *push* strategy, sellers introduce incentive programs (e.g., factory rebate, sales bonus, telemarketing, door-to-door sales, or discount coupons) to push products/services onto the consumers. Figure 4.22 illustrates the push strategy. Table 4.9 compares these two promotional strategies.

C. Promotion of High-Tech and Consumer Products/Services

High-tech and consumer products/services are promoted differently. To bring the most convincing marketing messages to the intended users,

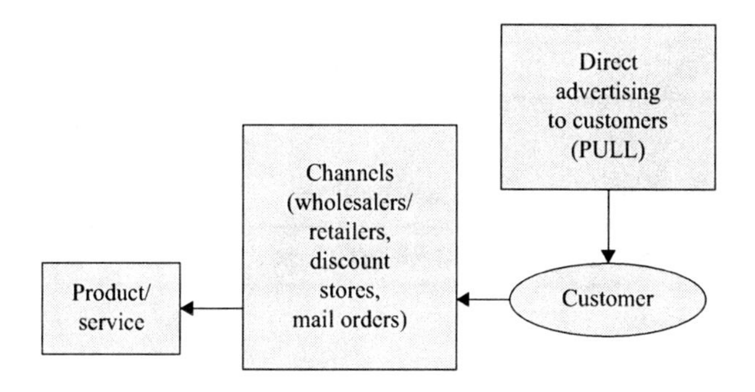

Figure 4.21. The pull strategy.

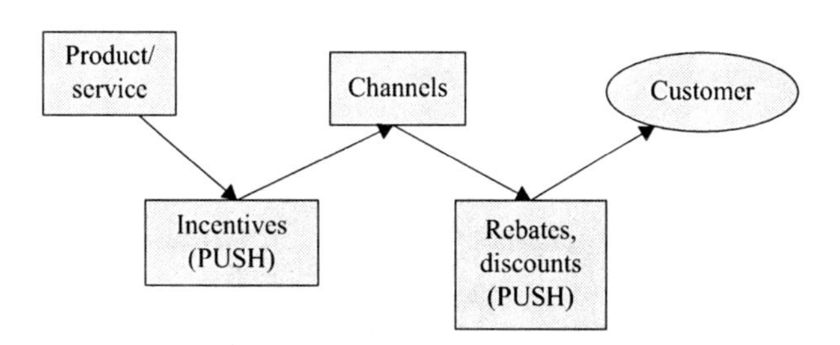

Figure 4.22. The push strategy.

marketers for high-tech and consumer products/services use different channels. Table 4.10 summarizes the major differences in tactics applied in marketing these products/services.

D. Internet-Enabled Communication Options

Communications among sellers, intermediaries (e.g., distribution partners), and buyers have been significantly enhanced by the Internet.[42] Figure 4.23 presents four specific modes of communication.

Manufacturers and suppliers usually set up the intranet to communicate with intermediaries (business to business, or B-to-B). Intermediaries may create their own websites and other tools to communicate with customers in a business-to-customer (B-to-C) mode. A direct communication

Table 4.9. Comparison of pull and push promotional strategies

Marketing factors	Push	Pull
Communication	Personal selling	Mass advertising
Price	High	Low
Service's need of special support	High	Low
Distribution	Selective	Broad

Table 4.10. Promotion of products/services

Promotion factors	High-tech offerings	Consumer offerings
Marketing costs	Low	High
Consumer segments	More	Less
Focus	More	Less
Advertising	Less important	More important
Marketing channels	Trade shows	TV
	User groups	Print media
	Trade journals	Internet
	Internet	Radio
Brand	Important	Critically important

Source: Ward, Light, and Goldstine (1999).

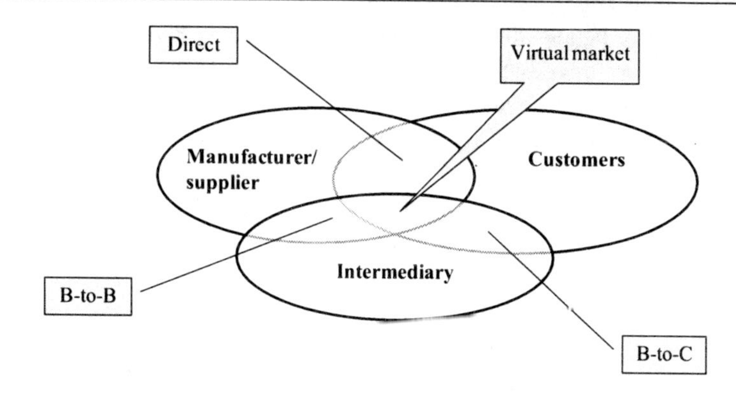

Figure 4.23. Modes of communications enhanced by the Internet
Source: Leverage Web for Corporate Success (1999).

among manufacturers, suppliers, and customers can be readily fostered by the company's websites, call centers, and other means for order processing, inquiry coordination, problem solving, and additional mission-oriented activities such as customer surveys, focus groups, and service testing.

The virtual market is a segment of the Internet domain wherein third-party portals (e.g., Google™, Yahoo®, and other search engines), auction sites (e.g., eBay®), and eMarketplace (e.g., ChemConnect®) actively provide channels to access information useful to all parties involved (e.g., manufacturers, suppliers, intermediaries, and customers).

In the B-to-B markets, businesses buy essentially two kinds of goods from other businesses: manufacturing inputs (raw materials, equipment, and components) and operational inputs (MRO goods, office supplies, spare parts, travel services, computer systems, cleaning, and others). They procure these goods and services by systematic sourcing or spot sourcing. Electronic hubs provide the useful functions of aggregation and matching.

E. Contextual Marketing

Companies invest efforts into creating websites that offer product/service information and facilitate sales transactions. However, studies show that these efforts have not yet returned the high profitability generally anticipated from such a marketing approach. The basic reason is that it remains unpredictable how frequently new and repeat customers visit these websites and then actually place orders there.

A new way of thinking is offered by Kenny and Marshall,[43] who suggest that the focus of e-commerce should be shifted from contents to context. They believe that contextual marketing (i.e., bringing the marketing message directly to the customer at the point of need) is the key. A number of contextual marketing examples are described next.

Johnson and Johnson's banner advertisements for Tylenol® (an over-the-counter pain-killing drug) show up on e-brokers' websites whenever the Dow Jones Industrial Average fails by more than 100 points on a given business day. They anticipate that investors will have headaches and thus will need Tylenol when they see their stocks lose money. The marketing message is brought out in the correct context as a way to reinforce its relevance and to offer transactional convenience at the right time.

CNET and ZDNET websites attract diverse visitors interested in computers. Instead of placing banner advertisements in these *CNET and ZDNET* sites to redirect visitors to its own website, *Dell* offers product information directly within the *CNET* and *ZDNET* websites. *Dell* piggybacks *on CNET's and ZDNET's* relationship with its computer-savvy customers in order to promote *Dell's* own customer-acquisition economics. Doing so holds the customer's attention on computer design and offers competitive design choices and speedy order processing at the optimal moment. Here, the tactic used by *Dell* is to insert itself into a preexisting customer relationship at the right time and place.

Several search engines (e.g., *Google, Yahoo, AOL*®, *MSN*®, *Lycos*®, *AltaVista*™, and *HotBot*) practice contextual marketing. When a user conducts a keyword search, the output is typically placed in a left-aligned column under the heading "Matching Sites" and rank ordered according to hit frequency. Frequently, several items under the heading "Sponsored Links" are placed on top of the "Matching Sites" column. These are paid advertisements related to the keywords entered by the user. They are there to offer contextual marketing messages relevant to the expressed interest of the user.

As Web-based technologies continue to advance, the Internet will become more accessible by many more users from almost anywhere and, at any time, causing them to become overwhelmed by information and choices. Bringing the right marketing information to the customer at the point of need is likely to become a critical success factor for various companies in marketing communications.

F. The Proper Approach to Communicate

In general, there are three ways to promote products or services: mass advertising, marketing to segment of one, and middle of the road marketing.[44]

1. **Mass advertising:** People resent being exposed to mass advertising, which often is unconnected to their lives or interests. In a recent survey, two-thirds of respondents said that they felt constantly bombarded by advertisements and 50 percent of them said that the advertisements they saw had little or no relevance to their lives. More than 60 percent of respondents look forward to new technologies that would block advertisements. At present, people use a number of tactics to avoid mass advertising. Examples include:
   (a) Use on-demand technology or digital video recorders to fast-forward through advertisements or to skip them completely.
   (b) Use mobile devices to download versions of popular shows free of commercials..
   (c) Internet users use software to block annoying spam mails and pop-up advertisements.
   (d) Use answering machines, caller ID, and the "Do Not Call" registry to avoid unsolicited phone interruptions at home.

   The future does not look good for mass advertising.
2. **Marketing to segment of one:** The other extreme is one-to-one marketing by offering customized products and services through targeted outreach to micro-segmented customers of high value. This "marketing to segment of one" approach requires a significant upfront investment, including (a) implementing customer relationship management software applications; (b) filtering, enhancing, and cleaning customer data; and (c) personalizing interactions—email, billing, offers, etc. These activities take time and coordination from multiple units of the organization (e.g., marketing, customer services, sales, IT), which can be daunting. What happens if the individual does not open the envelopes, pick up the phone or click on a box? The customers' concerns around privacy issues must also be taken into account.
3. **Middle of the road approach:** This approach targets broad groups of customers with messages that cannot be turned off. For example:
   (a) **Catch people in the bottlenecks in public space:** Target customers in the "bottlenecks," these are places where they cannot help but pay attention, such as riding in an elevator (wireless screens), a taxi (backseat video screen), an escalator (revolving stairs), flying from one city to another (in-flight movie channels, in-flight magazines, advertisements wrapped on airline tray tables), or using a public restroom (stall-mount messages). People see taxi rooftop screen showing nearby stores and restaurants, when the roving taxi is global positioning system

(GPS)-linked. The focus is on attracting the attention of an on-the-go but temporarily captive audience.

(b) **Use a Trojan Horse:** Advertise on frequently encountered materials such as paper coffee cups, paycheck stubs, pizza box advertising, take-out food containers, and garbage truck advertising. This advertising strategy is to infiltrate private spaces with mobile commercials.

(c) **Target people at play:** An example is to program the GPS of golf carts to alternatively show the golf course and selected advertising materials. The strategy is to reach people while they are pursuing leisure activities.

(d) **Get people to play games:** Display interactive posters in public spaces (e.g., malls, bus stations, and airports) so that customers can try out the products (e.g., MP3 players). The focus here is to engage people in ways that require them to interact physically with an advertisement or product marketing communication. Such a communication strategy would require a high degree of innovativeness to reach customers while purposefully challenging them.

Example 4.7

The company wishes to sell its current service to a new market segment. At the same time, it wants to launch a new service in the existing market segment. How should the company handle the service promotion?

Answer 4.7

To promote the company's current service in a new market segment, select advertisement media suitable for the new segment (e.g., services exhibitions, magazines, newspapers, etc.). To be emphasized are (a) favorable data such as customer satisfaction records, and response; (b) value addition; (c) current market share in other sectors; and (4) brand names established.

To launch a new service in an existing market segment, use current sales channels to cross-sell. Other steps include (a) training sales people, (b) preparing suitable sales literature, (c) setting up a service organization, (d) advertising in known mediums used in the past, and (e) offering incentives to push the new service.

4.5.4 PLACEMENT (DISTRIBUTION) STRATEGY

The *placement (distribution) strategy* defines issues such as (a) the product/service delivery options of an intensive, exclusive, or selective distribution system; (b) the company's relationship with dealers; and (c) the timely adjustments in distribution systems. Distribution assures that the right product/service is delivered at the right place and at the right time.

Numerous organizations are involved in moving products and services from the points of production to the points of consumption. As indicated in Figure 4.24, some companies may engage intermediaries (e.g., wholesalers and retailers) to distribute their products/services, while others may choose to interact directly with their customers.

Distribution channels serve a number of very useful functions, and these include: (a) *transportation*: overcoming the spatial gap between the producer and the users; (b) *inventory*: bridging the time gap between production and usage; (c) *allocation*: assigning quantity and lot size; (d) *assortment*: grouping compatible products/services for the convenience of users, since technical representatives may sell several service lines concurrently; (e) *financing*: facilitating timely possession of services; and (f) *communication*: providing product/service information to and feedback from consumers.

In recent years, distribution channels for some product/service vendors have experienced significant changes due to upgraded logistics, transportation technologies, and advancements in communication technologies. For example, the Internet has enabled many producers to deliver digitized services—books, newspapers, magazines, music programs, video games, and travel services—directly to consumers, thus bypassing the traditional intermediaries. Because of the use of sophisticated websites from which extensive product/service catalogs may be accessed, retail stores have

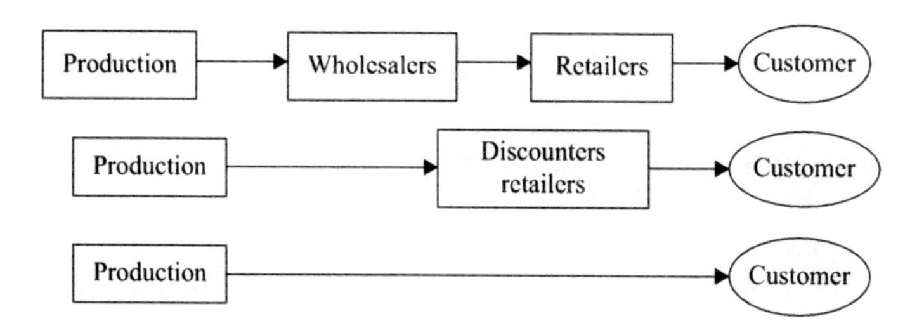

Figure 4.24. Distribution channels.

also lost some of their traditional importance in selling physical goods—clothing, cars, appliances, and so on.

Furthermore, logistic companies such as the *United Parcel Service (UPS)* are constantly improving their transportation capabilities and satellite-based communication system technologies in order to deliver physical goods anywhere in the world, while allowing their customers to constantly track the status of their orders.

Warehouse design is expected to increasingly involve gantry robots and complex process optimization for constantly improving the efficiency of automated high-volume operations.

A. Types of Distribution

Traditionally, distribution is classified as intensive, exclusive, or selective. In *intensive distribution*, products are stocked in diverse outlets, such as hardware stores, department stores, and catalog rooms, for wide distribution. This type of distribution is particularly suitable for consumer products of low technology and differentiation features. Examples include films, calculators, electric fans, books, and compact disks.

With *exclusive distribution*, certain products/services are distributed only through exclusively designated outlets. This allows producers to retain more control over price policy, promotion, credit, and service, as well as to enhance the image of the products/services. Examples include dealerships for specific cars and qualified service centers for brand-name investment services (e.g., Fidelity, Vanguard, etc.).

The *selective distribution* is suitable for certain other products, the sales and service of which require special technical know-how and training. Examples include electronic instruments, high-tech equipment, insurance brokerage, custom software, and others.

B. Organizational Structures

In order to enhance distribution effectiveness, some companies elect to exercise more control over the supply chain by integrating forward. Others have elected to integrate backward.

Forwardly integrated organizations strive to control the distribution channels leading to the customers. For the purpose of securing a larger market share and exercising more direct control, a producer may attempt to build his own retail outlets. Doing so allows the producer to gain a direct access to customers and thus benefit from their feedback.

On the other hand, *backwardly integrated* organizations seek to control the value chain leading backward to production. For example, some retailers or wholesalers may attempt to own specific production facilities or to outsource production for creating private-label products in order to market them with their own brand names, reduce costs, ensure supply, and control quality.

C. Impact of E-Commerce on Distribution

The Internet has significantly modified the traditional classifications of distribution. Many consumer services, as well as certain high-technology services, are now marketed directly through the company's websites, including order processing and after-sales services. As a consequence, many intermediate companies currently involved in distribution—wholesalers, discounters, and some retail stores—are gradually being forced out by the Internet-enabled e-commerce companies and by the increased involvement of efficient and fast-responding logistics providers.

One immediate impact of e-commerce is a possible reduction of the final product/service price and service delivery schedule, both of which are beneficial to end users.

Example 4.8

Customers' wants and needs are regionally different for products/services intended for global markets. How can a centralized, concurrent engineering team develop a product/service that will serve as the common "platform" for global markets?

Answer 4.8

Insofar as products are concerned, one option is to segregate the mechanical aspects (functionality) of the products from their aesthetic aspects (look and feel). *General Motors* is accomplishing this challenging objective by

A. Building identical assembly plants for Buick® cars at four global locations;
B. Outsourcing major subassemblies to local industries to lessen import duties and to satisfy local content laws;
C. Standardizing the technical specifications so that parts supplied by one region can be readily rerouted for use by another, in order to

balance loads due to market demand, labor disputes, governmental regulations, and other unpredictable events;

D. Modifying design to account for local market conditions relative to cultural preference (e.g., car names in local languages, styling preferences, purchase habits, colors, etc.);

E. Retaining centralized concurrent engineering approach to facilitate global business strategy and scale of economy, while being flexible enough to adjust to local needs.

With respect to services, the option could be to centrally develop and specify the core service element that contains novel and strategically differentiable features. This core service element is then supplemented by support elements (e.g., payment methods, delivery, information access, order processing, billing, exception, customer support, etc.), which are made locally adjustable.

Example 4.9

The company plans to enter a new global market. It has three services currently selling well in its home country. The company's brand name is strong and internationally well recognized.

Current market research indicates that the segments for these three services in the targeted global market are of different size, growth rate, and profitability for the foreseeable future. Other service characteristics are included in Table 4.11.

Which one service should be selected to penetrate the targeted global market? Why? If the company has the required resources to market all

Table 4.11. Product characteristics

	Service A	Service B	Service C
Segment size (dollars)	Small	Medium	High
Segment growth rate	Medium	Medium	Low
Profitability	High	High	Medium
Service value to customers	High	Medium	Medium
Brand strength	High	High	High
Delivery/distribution efficiency	Low	High	Medium
After-sales support activities	Medium	Medium	High

three services in the targeted global market, in what priority order should the company proceed?

Answer 4.9

To enter a global market, the company must examine two key questions: (a) How attractive is the target market segment to the company, and (b) how acceptable is the service offered to the customers in the target segment?

The attractiveness of a market segment to a company is generally defined by three factors: segment size, segment growth rate, and profitability. By using the information presented in Table 4.11, it becomes clear that the ranking based on "attractiveness" should be Service B first, with Service A and C sharing the second spot.

How acceptable the company's service is to the customers depends on the service value as perceived by the customers, the brand strength of the product, the delivery or distribution efficiency that affects the service's availability to the customers, and the ease with which customers obtain the needed after-sales support activities. Based on the "acceptability" criterion, the ranking of these products is Services B, C, and A.

Since both the "attractiveness" and "acceptability" criteria are equally important, we need to come up with a combination ranking, which says that the company should select Service B as its first choice to enter the global market, followed by Service C and then Service A.

Example 4.10

What are the bases for tradeoffs between conflicting wants and needs of different customers with respect to the same product? How important is it to emphasize product/service quality when a new and unique product/service is launched?

Answer 4.10

Customers make the following typical trade-offs:

1. Quality versus price;
2. Common features versus customization;
3. Automated self-service versus personalized attention;
4. Technical functionality versus styling and other aesthetic values

When launching a unique new product/service, quality is secondary to time to market and price. Thus, relatively speaking, it is not advisable to strive for high quality if doing so will delay the market entry and raise the product/service price. The strength of a unique new product/service lies in its novelty.

4.5.5 PHYSICAL EVIDENCE

The physical evidence refers to the physical setting (e.g., store front design, lobby appearance, layout and color, dress of service staff, ages of service equipment, print quality of service brochures, etc.) that has an impact on customer experience. Customers form a brand image of the product/service vendor by observing these physical evidence factors and by judging the extent to which these factors meet or exceed their expectations.

Apple is building a space-ship like headquarters at Cupertino, California, to promote its corporate image. *McDonald's and Citibank* design their retail branches to look and operate the same way in most locations.

4.5.6 PROCESS DESIGN

The process design specifies the applicable operations policies and work procedures to effectively serve the customers as related to order processing, logistics, inventory planning, franchising policy, sales training, procedures of delivering products/services, and the empowerment enjoyed by the customer-facing staff regarding problem solving. Customers gain an overall impression regarding the extent of the process being customer-focused.

4.5.7 PEOPLE

The people represents the key factor influencing customer experience, which is affected by the attitude, knowledge, and helpful and considerate behavior of the customer-facing service staff. Many people are directly or indirectly involved in the production and consumption of products/services (e.g., knowledge workers, employees, management, and other customers), who may add to the value of the product/service offerings. Following the Value Profit Chain model,[45] recruiting the right customer-facing staff, providing proper training in interpersonal skills, aptitude, and product/service knowledge, empowering them to take care of customers, and compensating them well are all essential strategies toward achieving customer satisfaction and corporate profitability.

To create favorable impression, companies also pay attention to dress codes. *Singapore Airlines* have maintained the same elegant uniform for its stewardesses for 25 years. Workers at *McKinsey & Company* are known to follow strict dress codes when meeting with business clients. *Aeroflot,* the Russian airlines, is known to have achieved great service improvement by having trained its flight attendants to memorize dialogs of pleasantries and reinforce rules on smiling, when interacting with customers, and to effectively replace the scouts, the cold shoulders, and the wordless encounters inherited from the traditional Russian culture.

Example 4.11

In this chapter, we talked about the marketing mix (seven Ps), which include (1) product/service, (2) price, (3) promotion (communications), (4) placement (distribution), (5) physical evidence, (6) process, and (7) people.

When marketing products, it is usually sufficient to focus on the first four of these seven marketing elements. On the other hand, all seven marketing elements are deemed important when marketing services. What are the underlying reasons for the last three (e.g., 5, 6, and 7) marketing elements to be particularly important for marketing services?

Answer 4.11

Companies address all seven marketing elements (seven Ps) when marketing services. The principal reason for this is that service (e.g., health care, business consultation, financial services, leisure and travel, insurance, and others) requires a much higher degree of customization in the process of specifying, producing, delivering, and offering after-sales services than products (e.g., automobiles, computers, appliances, etc.). Because of these service-specific characteristics, customers are exposed to the vendor's performance in (a) physical evidence, (b) process, and (c) people to a much greater extent, making these elements more important in affecting the customer's experience, than in marketing products.

Thus, to be customer focused, service companies pay more attention to (a) physical evidence related to facility design, office layout, and employee uniforms; (b) processes in problem solving and conflict resolutions affecting customers; and (c) people by choosing, training, and monitoring customer-facing staff to assure friendliness and customer-centered attention.

Example 4.12

A company makes a range of services and sells to several large, loyal customers to achieve a healthy market share. A new competitor has emerged to offer equivalent services at much lower prices. How should the company respond to this new threat?

Answer 4.12

The company has some short- and long-term options to deal with this potential threat to its market share position.

A. **Short-term options**
   1. Slash the service prices to a level lower than those offered by the new competitor for the sole purpose of driving the competitor out of the market. If the company's financial staying power is strong and the resolve is firm, the competitor may fold shortly thereafter.
   2. Establish a second brand—selling at the same price offered by the competitors so that there is no reason for the company's current loyal customers to switch. This strategy preserves the quality image of the first brand for the company.
   3. Add value by increasing customer relations management efforts to increase the switching costs (e.g., setting up Extranet, direct order line and programs, and special service personnel devoted to major customers).

B. **Long-term options**
   1. Revise service specifications to create differentiation in service features by closely listening to customers.
   2. Explore global sourcing opportunities to decrease cost, hence price, without diminishing service quality.

4.6 CUSTOMERS

Customers are important to any product/service company. Company's marketing program needs to focus on the targeted customer segments, understand them, practice the right techniques to acquire them, create emotional bond with them, improve interactions and enhance their loyalty, and continue to expand the number of satisfied customers by securing useful feedback.

4.6.1 CUSTOMER FOCUS

Customer focus is aimed at knowing the real needs of customers, in past, present, and future. It requires the collaboration of many employees as well as a functioning support organization to make it happen. Based on a study of the Royal Bank of Canada case,[46] a four-stage coordination process was suggested for service companies to become customer focused: (a) communal, (b) serial, (c) symbiotic, and (d) integral coordination; see Table 4.12.

Getting close to customers is a journey the entire company must take, not just the marketers and customer-facing staff. It requires corporate leadership and commitment to get useful results.

4.6.2 CUSTOMER ACQUISITION IN BUSINESS MARKETS

The benefits derived by gaining loyal customers may be classified into four categories.

1. Tangible financial benefits
2. Non-tangible financial benefits: conducting pilot projects, money-back guarantees for nonperformance, pay-for-performance contracts
3. Tangible nonfinancial benefits: corporate reputations, global scale, innovation capabilities,
4. Non-tangible nonfinancial benefits: something the vendor does extra for consumers to enhance convenience and customer relationship.

According to Narayandas,[47] companies must be on par with rivals on tangible financial benefits to acquire customers. They need to use tangible nonfinancial benefits to differentiate, shift customers' focus from tangible

Table 4.12. Process leading to customer focus

		Stage 1 Communal coordination	Stage 2 Serial coordination	Stage 3 Symbiotic coordination	Stage 4 Integral coordination
1	The primary organizational Objective	Collation of information	Gaining insight into customers from past behavior	Developing an understanding of likely future behavior	Real-time response to customers' needs
2	The coordination requirement	Communal coordination between a neutral information owner and the sources of customer information	Serial coordination among the neutral collator of information, analytics experts, and line organizations	Symbiotic coordination among the neutral collator of information, analytics experts, and line organization	Integral coordination among all of the company's employees across divisions, geographies, and other boundaries
3	The locus of leadership	Corporate strategy leaders and information technology	Corporate strategy leaders, the neutral entity that collates information (such as IT) analytics experts and marketing	Corporate leaders, customer segment managers or pivots, or both, that move information vertically and horizontally within the organization	Corporate leaders and cross-business integrators

Source: Gulati and Oldroyd (2005).

Table 4.13. Differences between consumer and business markets

No.	Characteristics	Consumer markets	Business markets
1	Segmentation of market branding	Critical	Not important—"segment of one"
		Important	Not important
	Focus of communication	Novelty of features	Solving customers' specific problems
2	Number of buyers	Large	Small
3	Transaction value	Small	Large
4	Production	Mass production	Customized to individual needs
5	Value	Customer's perception	Defined by customer's usage
6	Sales process	Brief	Elaborate (long and complex)
7	Retailing strategy	Important	Not important
8	Focus of sales efforts	End users	Group of decision makers

Source: Narayandas (2005).

benefits to nontangible nonfinancial ones (e.g., free services that reduce customers' operating expenses and suggest ways to make customers' process more efficient, thus earning the trust of the customers).

Consumer markets and business markets are quite different. Table 4.13 shows the difference.

When marketing to business customers, companies need to know that business decisions are typically made by a team of people. Each team member could have one or more specific needs. Knowing who these people are and what each of them is looking for is of critical importance to the vendor. The vendor must become sensitive to these needs and be prepared to address all of them adequately.

4.6.3 MOMENTS OF TRUTH IN CUSTOMER SERVICE

The "Moments of Truth" are time points when a customer invests a significant amount of emotional energy in the outcome of a product/service transaction with vendors. Examples of such moments include the following: (a) customer has a problem—encountering an unexpected difficulty in applications, receiving a hold on a check, and so on; (b) customer has a need to get a quick answer; and (c) customer receives financial advice (good or bad), and other advice. If these moments are handled positively, customers are likely to create an emotional bond with the vendor and subsequently increase their future commitment to the vendor's products/services. Studies[48] show that the customer experience related to other humdrum product/service interactions, which involve un-elevated emotional energy, is of little consequence.

Management must therefore support and develop front-line staff to enable them to handle such "moments of truth" by way of empowerment, nurturing the right service mindsets, and acquiring the needed service knowledge, while de-emphasizing the efficiency improvement of other humdrum transactions. Developing deeper and long-lasting relationships with customers is key to assuring long-term profitability.[49]

4.6.4 CUSTOMER INTERACTIONS AND LOYALTY

The interactions between customers and companies play a very important role in securing marketing success. Creating a pleasant experience for customers (e.g., in order processing, service information dissimilation, inquiry coordination, problem solving, after-sales service, and market surveys) is crucial for customer retention. Winning customer cooperation in offering much-needed feedback is vital to the company's new product/service development. Management of customer relations is thus an important corporate responsibility.

Some lessons from the past are noteworthy. Customer interactions are not limited to marketing. Many other functions of the company are involved, including product/service design, accounts receivable, legal, engineering, manufacturing, and shipping. Empowered employees can act on behalf of the company to satisfy customer requirements. Adequate support infrastructure must be established to enable employees to perform these tasks in an innovative and customer-responsive manner.

The major payoff of a successful customer interaction program is customer loyalty. Customer loyalty contributes to company profitability. Studies indicate that increasing customer retention rate by 5 percent could raise profits by 25 to 95 percent.[50] Loyal customers are valuable because they buy more, refer their pleasant experience to new customers, and offer consultations to these new customers at no cost to companies.

To build customer loyalty, the customer interaction strategy must be focused on creating trust. *Amazon.com* is viewed by many as one of the most reliable and trustworthy websites on the Internet. It registers user preference, becomes smarter over time at offering products/services tailored to each user, provides one-click convenience for purchasing items, and delivers the ordered products/services free of errors. It is reported that 59 percent of *Amazon.com* sales are derived from repeat customers, roughly twice the rate of typical "bricks-and-mortar" bookstores.

Vanguard Group, a company that markets index-based mutual funds, offers timely and high-quality financial advice on its website and does not attempt to hard sell any specific service. Its customer interaction strategy is focused on building trust. "You cannot buy trust with advertising or salesmanship. You have to earn it by always acting in the best interest of customers," says Jack Brennan, Vanguard chief executive officer.

eBay is known to have over 50 percent of its new customers referred by loyal customers who also serve as helpers to them. One major concern in the business of auctioning used merchandise is reliability and fraud prevention. *eBay* asks each buyer and seller to rate each other after every transaction. The ratings are posted on the website. Every member's reputation becomes public record. Furthermore, *eBay* insures the first $200 for each transaction and holds the money in escrow until the buyer is satisfied with the received product.

How is trust related to profitability? Studies show that, in some businesses, customers must typically stay on board for at least 2 to 3 years just for the companies to recoup their initial customer-acquisition investment. In other words, for companies to achieve profitability, customers must be loyal enough to stay beyond this break-even period. A large percentage of customers defect before many new companies reach this break-even point. Table 4.14 lists statistics related to customer acquisition cost, years to break even, and percentage of customers who defect before the break-even point.

A large number of companies are successful in planning and implementing strategies to interact effectively with customers. These companies identify and prioritize customers, define their needs, and customize products/services to fit these needs.[51] They reap the benefits of increased cross-selling, reduced customer attrition, enhanced customer satisfaction, minimized transaction costs, and sped-up cycle times. Companies known for their success in relationship marketing include *Pitney Bowes, Wells Fargo, 3M, Owens Corning, British Airways, Hewlett-Packard, and American Express.*

Customer loyalty is won, not by technology, but through the delivery of a consistently superior customer experience. It requires a well-designed customer interaction strategy that is supported by companies with a firm corporate commitment.

Table 4.14. Customer-loyalty-related statistics

Products	Acquisition cost per customer (dollar)	Years to break even	Percentage of customers defecting before break-even points (%)
Consumer electronics and appliances	56	4+	60+
Groceries	84	1.7	40
Apparel	53	1.1	15

Source: Reichhold and Shelfter (2000).[50]

4.6.5 CUSTOMER FEEDBACK—THE ULTIMATE QUESTION

A great number of customer-satisfaction surveys contain too many questions. The ultimate question to ask is: "How likely are you to recommend this company to a friend or colleague?"[52] Score the results in a 0 to 10 scale and classify the responses as follows:

1. Loyal promoters (9s and 10s)
2. Customers who do business passively with the company (7s and 8s).
3. Detractors (6s and below)

The net promoter score (NPS) is defined as the percentage of promoters minus the percentage of detractors. For example, if the promoters are 35 percent, passive customers are 50 percent, and detractors are 15 percent, then the NPS is 20 percent. Based on a survey conducted by Bain & Company, companies with NPS in the range of 50 to 80 percent, are superior in achieving good profits; see Table 4.15.

The key to growing a business is to have more promoters and less detractors. Since the company already has a relationship with these "promoter" customers, it should solicit information from these customers in order to understand exactly where the company is succeeding and how to apply this information toward ensuring the company's ongoing success. It should contact detractors as well, to find out what the company can do to improve its service offerings. Reihheld[53] believes that business growth can only be sustained over a long period based on the loyalty of satisfied customers and customer-initiated promotions. To achieve success, companies must always practice the golden rule: to treat others as they would want to be treated in return.

Table 4.15. NPS scores of selected U.S. companies

Selected companies	NPS score (%)
USAA	82
HomeBanc	81
Harley-Davidson	81
Costco	79
Amazon.com	73
Chick-fill-A	72
eBay	71
Vanguard	70
SAS	66
Apple	66
Intuit (TurboTax)	58
Cisco	57
FedEx	56
Southwest Airlines	51
American Express	50
Commerce Bank	50
Dell	50
Adobe	48
Electronic Arts	48

Source: Reichheld (2006).

This concept appears to be consistent with the "The Value Profit Chain" model (Section 4.5.6) in that satisfied customers will not only increase spending for themselves, but also recommend that their friends follow suit. The way to ensure customer satisfaction is to adopt company policies that make for happy and loyal employees, who will in turn provide excellent customer service leading to long-term corporate profitability.

Example 4.13

Over the years, Company XYZ spent a considerable amount of efforts in developing and testing a new drug intended for reducing the low-density lipoprotein (LDL: bad cholesterol) and raising high-density lipoprotein

(HDL: good cholesterol) of patients with cardiovascular disease. After having passed phases 1, 2, and 3 trials, the drug received Food and Drug Administration (FDA) approval for marketing to the public. There are some known drugs already in the marketplace for this type of heart disease, which affect millions of people in the United States alone. The size of the overall market for cardiovascular drugs is estimated to be about $25 billion annually. Devise a marketing plan to bring this new drug into the U.S. marketplace.

Answer 4.13

The marketing plan should consist of four parts, corresponding to the four Ps of marketing products, namely, product, price, placement (distribution), and promotion.

As a new product, the drug in question offers the useful features of lowering LDL and raising HDL, a very powerful combination to combat heart disease, based on the current state of clinical knowledge. A direct competitor is Lipitor, which has the same combination effect as the new drug. Furthermore, Crestor, together with Niacin, is also known to produce this combination effect. It is important for the company to delineate any differences this new drug might have regarding (a) seriousness of any side effects, (b) potential of long-term health hazards, (c) lower frequency of taking the drug, and (d) longer effectiveness. These differences must be clinically verifiable by ways of large-scale clinical studies.

Pricing is an important issue to some patients. The new drug should be retail-priced at a level slightly lower than that of its current competitors. Aggressive contracts should be entered with major insurance carriers, mail order drug companies, American Association of Retired Persons (AARP) prescription drugs program, and others to allow volume-based discounts.

Promotion is rather critical for the new drug entering an existing market. Key targets are physicians, patients, and insurance carriers. Patients need to become aware of the unique benefits of this new drug via TV, magazine articles, and web-based advertisement, so that they could "pull" this drug from the supply chains. The message should emphasize its distinguishing features in view of the existing competition. Physicians must be convinced via trade shows, clinical studies, and publications of its merits, so that they would be willing to suggest/prescribe this new drug and allow brochures to be distributed through their offices. Frequent publication of supporting articles in highly reputable journals such as *New England Journal of Medicine*, *Journal of American Medical Association*, *Journal of Cardiology*, *Circulation*, and others will garner the attention of physicians. Insurance carriers need to be convinced of the benefits of this new drug in order for them to be willing to place it in their formularies.

Distribution of this new drug would follow the usual wholesale channels, as it is a prescription drug authorized by physicians and available only from pharmacies. Company XYZ should monitor the market constantly and adjust its marketing program accordingly.

4.7 OTHER FACTORS AFFECTING MARKETING SUCCESS

There are several other factors that may affect the marketing success of any company.

4.7.1 *ALLIANCES AND PARTNERSHIPS*

Nowadays, companies realize increasingly that they do not always have, or cannot cultivate internally, all competencies needed to compete in the world markets with the resource and time constraints under which they have to work. Because market access may be unattainable, the technology unaffordable, resources unavailable, time to market too long, or because other barriers may exist, an individual company may find it increasingly difficult to compete alone. Partnerships and alliances have become more and more necessary for some companies to compete effectually and to constantly deliver value to customers.

For companies to succeed in the marketplace under these circumstances, the marketing concepts must penetrate to all members of the partnerships and alliances.[54] All partners must appreciate that mutual gain will result only when all members of the alliance embrace the marketing concept and come to recognize the importance of creating superior customer value by joining hands.

4.7.2 *ORGANIZATIONAL EFFECTIVENESS*

Marketing success is influenced by how effectively the company operates. In general, organizations with less rigid structure have a higher likelihood of becoming more customer focused, technologically innovative, and market responsive. Certainly, any conflicts among internal functions (e.g., manufacturing, design, engineering, and marketing) must be minimized. Technology for mass customization requires an integration of

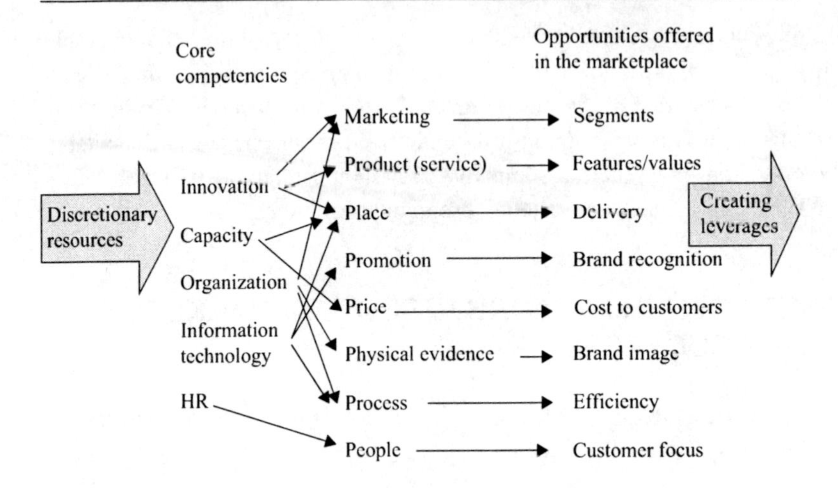

Figure 4.25. Organizational effectiveness.

R&D, procurement, customer relations management, and supply-chain management to achieve a high degree of customer satisfaction.

Above all, company management must apply discretionary resources (e.g., R&D, production capacity, human resources, organizational expertise, and information services) to the right combination of strategies (e.g., marketing, service, distribution, promotion, and price) so that maximum strategic marketing leverage can be achieved to capture opportunities offered in the marketplace. Figure 4.25 illustrates this core concept of organizational effectiveness.

Example 4.14

Engineering refined the design specifications of a product/service as originally recommended by marketing. Manufacturing made further changes to the service design in order to fabricate the product/service automatically. Unfortunately, the product/service did not sell in sufficient quantities to make it a success. Explain the possible reasons.

Answer 4.14

The product/service features defined initially by the marketing department may not be exactly what the majority of customers wants. Product/service testing is a critical step to fine-tune the design. The selected method of production does not readily accommodate an adjustment of product/service features, even if they are identified by feedback from the marketplace. The manufacturing of the product/service should be based on demand to assure market acceptance, not based on production technology, which is aimed at cost reduction.

Skipping the product/service testing step and applying the automatic production method too soon are two likely reasons for the noted failure.

Example 4.15

Organizational effectiveness is necessary for any corporate leader to attain company leverage and hence strategic differentiation in the marketplace. Explain in your own words how corporate leaders should go about realizing this much needed organizational effectiveness?

Answer 4.15

Organizational effectiveness is the degree to which company managers are in a position to optimally apply core competencies (e.g., innovations, capacity, organization, IT, and people—both internal and external) in driving the seven Ps of marketing (product/service, pricing, placement, promotion, physical evidence, process and people) to successfully capture opportunities in the marketplace and gain leverages; see Figure 4.25.

They need to nurture these core competencies (providing incentives for innovative and creative outputs), mobilize resources (e.g., both internal and external talents), understand customers' needs (e.g., with an open mind), achieve process excellence (e.g., applying known IT and other tools to simplify and assure the quality of customer interactions), and adapt to market changes, in order to create strategic differentiations and sustainable profitability.

4.8 CONCLUSIONS

This chapter covers many important issues related to the marketing of the company's products and services. Engineering managers should understand the overall objectives of the firm's marketing efforts and become sensitive to various marketing issues affecting engineering. They should become well versed with marketing terminology and elements of the marketing mix, namely, product (service), price, promotion, distribution, physical evidence, process, and people. It is important for engineering managers to accept the fact that marketing of products/services involves a lot of uncertainties associated with consumers' perceptions, competitive analyses, market forecasting, and people-related issues. They need to wholeheartedly adopt a customer orientation in planning and implementing all engineering programs. They must strive to work closely with marketing personnel and remain supportive of the overall marketing efforts by providing high-quality engineering/technology inputs to the firm's marketing program. Having a penchant for doing this is a key for them to succeed in industry.

Obviously, the engineering/technology inputs most useful to marketing are related to products/services and associated production and delivery activities. These include specifying and designing innovative and differentiable product/service features to be of value to customers, utilizing technologies to confer competitive advantages in time to market, quality, reliability, and convenience, and delivering after-sales support activities needed to ensure customer satisfaction.

Engineers are also expected to control costs by improving and managing the production process, resource (labor and materials) allocation, and quality control. They may also get involved in training sales people, making presentations before customers, conducting industrial exhibits, and evaluating customer feedback related to new service features.

Having learnt the marketing concepts and been exposed to the complex marketing issues reviewed in this chapter, engineering managers and professionals will be able to appreciate the difficult but critically important functions of marketing and can become more effectual in interacting with marketing management.

Marketing and innovations are two principal activities of any profit-seeking organization. Engineers already know how to innovate. If they also learn how to market, this combination of capabilities will surely enable them to become major contributors in any product/service organization.

4.9 APPENDICES

4.9.1 CONSUMER SURVEY AND MARKET RESEARCH

To market consumer products, companies need to have a very detailed understanding of their customers, just as companies marketing industrial products also need to understand their industrial customers, although to a much lesser extent.

When dealing with customers, the key questions typically concern what, how, where, when, why, and who. For example, what functions does the product serve? What are the criteria to buy the product (price, color, size)? What is the value to the customer? What do they really want from the product (psychological, functional, and other benefits)?

How do customers compare products? How do customers decide to buy? How is the product to be used? How much are customers willing to pay for it? How much do they buy? How would the distribution mode and service center location affect the customers' buying decision?

Where is the purchase decision made (e.g., what is the customer's position in the company or household)? Where do they receive information from? Where do they buy their products (e.g., retail store, mail order, department store, Internet, etc.)?

When do they buy it (weekly, monthly, special occasions)?

Why do they prefer one brand over the other (e.g., performance, price, convenience, packaging, colors, service, etc.)?

Who are the customers (e.g., age, background, sex, geographic location, members of social groups, etc.)? Who buys the competitor's products? Who does the buying (wife, husband, children, purchasing agent, engineers, others)? Who makes what decision for whom (decision-making entities)?

To understand the behavior of consumers in making purchase decisions, companies focus on customers' habit in purchasing, consuming, and information gathering. Who buys, how often, where, how much, when, and at what price? Who consumes, on what occasions, how do they consume, in what quantities, where, when, and with what other products? What media do they use (industrial exhibits, trade journals, TV, newspaper, radio, Internet, etc.), and when?

It is also useful for companies to understand the process by which consumers make their purchase decisions. This process typically encompasses the steps of need arousal (e.g., problems to solve; discovery from neutral sources; Jones the bragging next-door neighbor; etc.); information search (online resources are now one click away); and evaluation and decision making (e.g., comparative shopping, making trade-offs, brand versus product, and price versus quality and features).

CHAPTER 5

Conclusions

This book covers a set of business fundamentals that are particularly useful to engineering managers and professionals. Cather[1] suggested a large set of business skills for engineers and technologies, including project management, human resource management, customer needs, laws, IT, and e-commerce. While some of these skills are of general benefit and useful, the three sets of business skills included in this book are the most impactful to the work involving technically talented engineering managers and professionals.

Leadership is built on superior vision. Vision defines business directions in anticipating the future conditions, which in turn are affected by technology, marketplace, customers, resources, regional economics, and others. Vision is the capability of seeing into the future. Without the benefits of having business fundamentals, the future one can see is rather narrow. Jack Welch said: "Good business leaders create a vision, articulate the vision, passionately own the vision, and relentlessly drive it to completion." Engineering managers and professionals need to demonstrate an ability to set aside a parochial technology mind-set and look at the broader picture. They will need to paint the future on a wide canvas. The future has several dimensions, such as marketplace, customers, financial metrics (e.g., growth, gross margin, return on assets), production, and IT, with technologies being just one of them. Broad-based visions enable them to conceptualize the future, identify unstoppable trends, and develop new ways to grow. Such broad-based visions are needed to lead at senior management levels. Business fundamentals will help engineering managers and professionals to move into the senior management positions and perform successfully there. Good leaders light a fire under us to get things done.

Of critical importance is for engineering managers and professionals to constantly strive for contributing significantly to help achieving (a) strategic differentiations (i.e., by way of creating novel products or services to be ahead of the competition in winning the acceptance of demanding customers) and (b) operational excellence (i.e., through constantly improved work processes to produce and deliver products and services). Such valuc-adding contributions can be made much more often, when supported by skills and knowledge in cost accounting, financial account and analysis, and marketing development. Only products/services that offer novel and distinctive features, are cost-effectively produced, meet customers' needs, create pleasant experience to them, sell large enough units, and generate significant profitability, will be of lasting value to an enterprise.

For engineering managers and professionals to advance in a highly competitive business environment, they need to be able to communicate efficiently, using the right business vocabulary and business insights, with other leaders in the organization. As most major decisions in a given enterprise are business-related, the issues of cost, financial metrics, marketing, and customers become inherently important. Having acquired the business perspectives and being able to proactively initiate new projects, engineering managers and professionals will gain opportunity to offer quality judgments and broad-based decisions that garner support from other corporate decision makers. As Aristotle Onassis said: "To succeed in business it is necessary to make others see things as you see them."

As engineering managers and professionals continue gain influence by the quality of their judgments and decisions, which are greatly enhanced by their technological and business perspectives, they will likely be favored for joining the senior management ranks. Progressive companies are constantly developing future leaders with broad experience in design, engineering, marketing, and finance. As it is in practice with many Fortune 500 enterprises, many of these up-and-coming engineering managers are further being supported for executive type of business training (e.g., Executive MBA programs, which meet on Saturdays), to ready them for higher-level leadership positions. The importance of being well versed in marketing, finance, and costing is rather obvious.

For engineering managers and professionals to be successful, they will need not only to acquire business management fundamental, but also to practice such knowledge and insights in order to derive real benefits for the enterprise. Anton Chekhow said: "Knowledge is of no value unless you put it into practice." A secret of success for them is to initiate technically novel projects that deliver real value to their enterprises. Proactively pursuing important projects and implementing them successfully are vital for the long-term success of engineering managers and professionals. Examples of such projects could include: (a) new product/service development, (b) new growth opportunity assessment, (c) competitive responses to external threats, (d) productivity improvement, (e) process enhancement, (f) value engineering, (g) new technology selection and incorporation (e.g., cloud-based, mobile), (h) development of new business models, (i) creation of new external alliances (marketing, supply, co-development of technologies), (j) cost reduction and quality control, (k) knowledge management initiatives, and others. Patty Hansen said: "You create your opportunity by asking for them."

Many of the above-described projects are likely to involve teams of highly qualified specialists, who are to be managed by leaders with broad technical and business backgrounds. Teams are organizational forms, which promote "Together everyone achieves more." Woodrow Wilson suggested: "Not only use all the brains that I have, but all that I can borrow." For the example of developing new products/services, the project teams are generally expected to perform a large number of sequential tasks, following a typical state-gate process:[2]

1. **Define project objective** (specify the type of new product/service to develop, novelty of the project idea from the patentability and competitive standpoints, major features that deliver values to customers, completion date, the estimated investment budget, alignment of project outcome with enterprise's business objective, etc.) The new features so defined should vault the company ahead of its competition.
2. **Conduct marketing research** (understand customer needs, existing and future competitions, size of the customer segments, market growth rate, governmental rules and regulations, etc.)
3. **Assess financial feasibility** (create a multiyear income statement, define product/service life, estimate additional charges for corporate overhead, such as administrative and corporate depreciation, etc. Compute estimated annual cash flow and define the net present value [NPV] for the project).
4. **Determine technical feasibility** (production capability, technical know-how, supplier capabilities, outside business partners, etc.)
5. **Secure project approval** (gain authorization to proceed, including budget, resources, technologies, and management attention)
6. **Specify major stages and gates** (identify stages in the development process, and the goals for each, including performance metrics that define progress, and review results to assure a continued success)

7. **Formulate project team** (define team leader, suitable members for each stage involved, roles and responsibilities of each, communication rules, technologies to be used, etc.).The team leader's dedication toward the project will inculcate a sense of responsibility and urgency among the rest of the team.
8. **Design products/services** (introduce novel functionality and features, paying more heed to customers, the use of new technology, etc.)
9. **Calculate unit costs** (material, labor, overhead, capital investment, staff training costs, equipment maintenance, depreciation, energy usage, unit cost of production, warehousing, distribution, and others). Apply activity-based costing, if needed.
10. **Produce, test and improve prototypes** (make prototypes, perform value engineering, define outside partners and suppliers, conduct testing in selected regions for functionalities and degree of customer acceptance, solicit user's feedback, and refine design, etc.)
11. **Activate marketing program** (define price, specify advertising channels, estimate future sales, formulate customer service procedures, secure customer feedback, engage outside partners, etc.)
12. **Achieve profitability and preserve learning** (set up suitable teams to implement all steps needed to gain profitability, carefully preserve team knowledge and insights, and reuse the know-how to constantly enhance corporate competitiveness)

Business fundamentals are essential for carrying out the steps: 1, 2, 3, 5, 8, 9, and 11. Figure 5.1 shows the known stage–gate process commonly used by many enterprises to achieve project success.

As an example, Figure 5.1 illustrates a total of six stages. The project moves forward, only if all objectives in a given stage are met. New team members may be added in the following stages, dependent on the stage's principal emphasis under development. The size of the hexagons in Figure 5.1 becomes smaller as the process moves forward, indicating that the unresolved issues become lesser in number. The last stage (i.e., lessons learned and reuse) is important from the knowledge management standpoint, as the enterprise needs to acquire new knowledge and insights to remain competitive over time.

A critical hurdle to overcome in all project development processes is the one involving management approval, which authorizes the needed budget, manpower, and management priority. To secure such a project

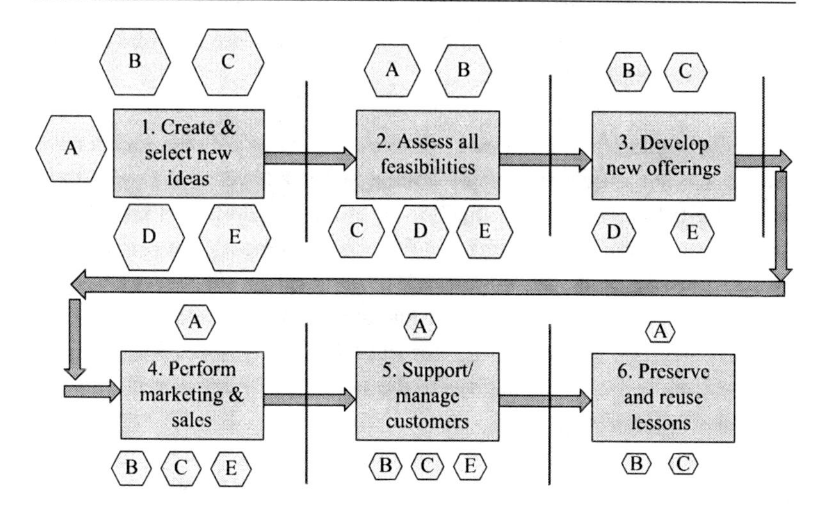

Figure 5.1. The stage–gate process of developing new products/services Innovations .

Notes: (A) Business management (strategic alignment, new idea selection, and patentability review)

(B) Technical and financial feasibility assessment

(C) Marketplace acceptance

(D) Production

(E) Customer support/management

approval, engineering managers and professionals need to be familiar with the questions that are typically asked by senior-level decision makers:

1. How novel is the new product/service, from the viewpoints of its patentability and the existing competition?
2. In what specific ways will customers benefit from the use of this novel product/service? How important is this value proposition to them?
3. How large (in annual sales) might the estimated market size be for such a product/service in five years?
4. How technically feasible is it to be produced? Are current technologies sufficient for its production? (see Figure 5.1)
5. How much will the initial investment be, as well as in subsequent years?
6. What is its NPV, after the product/service life runs out? The NPV of the project is the net value added to the enterprise, after having paid off all initial investment, design, production, marketing, sales,

customer services, and other expenses, which are required for the project implementation.

If the engineering manager or professional, who initiates such a new product/service proposal, cannot convincingly answer these questions (including the reasonable justification of all key assumptions introduced during the creation of the multiyear financial statement), the project will not be approved. Thus, the importance of engineering managers and professionals to acquire some business fundamentals discussed in this book is clearly evident, as they would need just to score a few of such high-NPV projects to propel themselves to the senior management ranks and to assure their places there.

At the senior management levels, leaders think more in strategic terms. Raynor and Ahmed[3] studied the performance of 25,000 companies from 1966 to 2010 and discovered the key factors, which enabled 344 excellent companies, including *Merck*, *Maytag*, and *Abercrombe& Fitch*, to sustain business success. These excellent companies make decision and execute strategies following three rules: (a) better before cheaper, (b) revenue before cost, and (c) use nothing but only these two rules.

1. **Better before cheaper:** To be better, the products and services need to be of high quality and include distinguishable functionalities and features that deliver value to customers. In addition, activities that add nonprice benefits to customers are also to be emphasized, such as customer support, problem solving, brand name, and customer relationship management. To be emphasized to a lesser extent is the product/service price. This strategy says essentially that companies should favor premium products/services, rather than commodity type, which would compete on price only in the marketplace. In other words, companies will do well over the long term if they emphasize strategic differentiation, by coming up with premium products/service, which allow them to charge high prices and preemptively market novel offerings. A high-capitalization company that follows this rule is *Apple*. Peter Drucker said: "Quality in a product or service is what customer gets out and is willing to pay for. Nothing else constitutes quality."
2. **Revenue before cost:** Focusing on revenue means paying attention to generate the top-line sales revenue by expanding the market reach. Cost can be reduced using various operational excellence approaches, such as Lean Six Sigma, Value Stream Mapping, and Failure Mode

and Effect Analysis. Generally speaking, cost reduction has natural limits, beyond which the marginal return of such efforts will diminish. Again this strategy favors the acquisition of sale revenue using new and novel products/services, which could be sold in new markets, over the continued drive for cost cutting, thus favoring the pursuit of strategic differentiation over that of operational excellence.

Following the above-described rules requires business leaders to make use of their extensive knowledge regarding costs, financial accounting, and market management. Those engineering managers and professionals who have acquired the business fundamentals contained in this book are in a better position to think in such strategic ways. Those who can actively contribute to achieve strategic differentiation and, to a lesser extent, also operational excellence for their employers will be among the progressive leaders, whom we need more to practice the strategies of "better before cheaper" and "revenue before cost" and effectively guide excellent business enterprises.

Keeping the products and services cheap requires a good understanding of costs, the ways these costs are estimated, and how they can be constantly improved. Revenue is raised by expanding the customer base and defining a unit price that is attractive in view of the the competition.

In a typical college curriculum, cost accounting, financial account and analysis, and marking management could each be easily a 15-week-semester-long course. Even though the parts of business fundamentals that are most useful to professionals in science, technology, engineering and mathematics (STEM) disciplines are already included in this book, it is still useful for engineering managers and professionals to get exposure to other details by browsing through some of the basic textbooks on these subjects, if needed.[4]

In conclusion, the business fundamentals discussed in this book could assist engineering managers and professionals to accomplish quite a few objectives. Thus, the takeaway of this book may be summarized as follows:

1. These business fundamentals are most impactful to work done by technically talented people in for-profit enterprises.
2. They are essential for them to exert leadership in many for-profit organizations.
3. They will enable them to communicate more effectively with important peers and decision makers in the organization, using pertinent business vocabulary and broad-based business insights.

4. They are likely to prepare them well for advancing into senior management ranks and enable them to develop the needed business vision to stay there successfully. Benjamin Franklin said: "By failing to prepare, you are preparing to fail."
5. They will enable them to contribute to the creation of strategic differentiation and operational excellence of the enterprise, such as initiating and successfully implementing technology-based projects, which add measurable value and confer business advantages. Not focusing on creating strategic differentiation and operational excellence may lead to the quite common observation: "The harder we work, the behinder we get."
6. They will enable them to think and act like progressive leaders in excellent business enterprises, wherein the forward-looking strategies of "better over cheaper" and "revenue over cost" are actively pursued.

David Starr Jordan said: "Wisdom is knowing what to do next, virtue is doing it." Engineering managers and professionals need to learn constantly and take proper actions regularly to add value. They are too good (in science, technology, engineering, and math), not to be better by mastering additional business fundamentals.

NOTES

Chapter 1

1. Coplin (2012).
2. Cather et al. (2001).
3. Brown (2010).
4. McCubbrey (2001).
5. Babson (2005).

Chapter 2

1. Lanen, Anderson, and Maher (2013); Horngren, Foster, Datar, and Rajan (2011).
2. Bragg (2002); Rapier (1996).
3. Wiese (2013).
4. Cokins (2006); Sanford (2011); Moore (2012).
5. Wiese (2013).
6. Atkinson et al. (2007).
7. Bamber and Hughes (2001).
8. Waters et al. (2003).
9. Buttross and Schmelzle (2008).
10. Waters et al (2003).
11. Briner, Alford, and Noble (2003).
12. Fichman and Kemerer (2002).
13. Swenson, Ansari, and Kim (2003).
14. Frenkel et al. (2013); Cox (2010).
15. Akira (2009).
16. Mun (2010); Wang (2012); Smith (2011).
17. Charnes (2012).
18. Charnes (2012).
19. Canada et al. (2004).
20. Touran (2003).
21. Baker and English (2011); Sullivan, Wicks, and Koelling (2011).

22. Nguyen and Walker (2005); Ross (2004).
23. Reznik and Dimitroy (2013).
24. Nimocks (2005).
25. Touran (2003).
26. Horngren, Foster, Datar, and Rajan (2011).

Chapter 3

1. Kimmel, Waygandt, and Kieso (2012).
2. Sydsaeter and Hammond (2012).
3. Ross, Westerfield, and Jordan (2012); Fields (2011).
4. Brigham and Ehrhardt (2013); Brealey, Myers, and Alen (2013).
5. Droms (2003).
6. Berman, Knight, and Case (2013); Revsine et al. (2011).
7. Skonieczny (2012).
8. Berman,Knight, and Case (2013); Ittelson (2009).
9. Makoujv (2010).
10. Healy and Choudhary (2001A).
11. Healy and Choudhary (2001A).
12. Healy and Choudhary (2001B).
13. Healy (2001).
14. Healy and Choudhary (2001C)
15. Subramanyam and Wild (2013).
16. Skonieczny (2012).
17. Healy and Cohen (2000); Troy (2012).
18. Standard & Poor's (2012).
19. Bragg (2012).
20. Ferri, Ferris, Treadwell, and Desai (2006).
21. Rappaport (2006).
22. Delong et al. (2005).
23. Tregoe and Kepner (2006).
24. Kaplan and Norton (2007).
25. Kaplan and Norton (2007).
26. Kaplan and Norton (2007).
27. Catucci (2003).
28. Luehrman (1997).
29. Kelleher and MacCormack (2005).
30. Reeve et al. (2011).
31. Charnes (2012); Wang (2012).
32. Luerhman (2009).
33. Black and Scholes (1973).
34. Lacey and Chambers (2010); Titman and Martin (2011); Koller, Goedhard, and Wessels (2010).
35. Walker (2013).

Chapter 4

1. Kotler and Keller (2011); Chemey and Kotler (2012).
2. French and Knudsen (2007).
3. Lovelock and Wirtz (2010); Palmer (2011); Baron et al. (2009); Greimer et al. (2012); Wirtz et al. (2012).
4. Griffin (2002).
5. Smith and Kawasaki (2011).
6. Schultz et al. (2013); Waugh (2004).
7. Jones (2006).
8. Ehrlich and Fanelli (2012); Ennew and Waite (2013).
9. Glass (2012).
10. Strouse (2004).
11. Kotler et al. (2013).
12. Thomas (2009).
13. Fitzpatrick (2013).
14. Keegan (2013).
15. Ferrell and Hartline (2010).
16. Barnett (2009).
17. Grenci and Watts (2007).
18. Hartley and Claycomb (2013).
19. Yankelovich and Meer (2006).
20. Yankelovich and Meer (2006).
21. Yankelovich and Meer (2006).
22. Hoffman and Bateson (2010)
23. Edvardsson et al. (2006).
24. Miller and Palmer (2000); Marks et al. (2012).
25. Lovelock and Wirtz (2010).
26. Stark (2011); Moon (2005).
27. Neufeld (2005).
28. Dev (2008).
29. Keller (2000).
30. Ward, Light, and Goldstine (1999).
31. Keller, Sternthal, and Tybout (2002).
32. Bedbury with Fenichell (2003).
33. Wheeler (2012).
34. Berry and Seltman (2007).
35. Berry and Seltman (2007).
36. Rust et al. (2004).
37. Rust et al. (2004).
38. Dolan and Gourville (2005).
39. Smock (2003).
40. Chaffey et al. (2009).
41. Gourville and Soman (2002).
42. Roberts and Zahay (2012).

43. Kenny and Marshall (2000).
44. Nunes and Merrihue(2007).
45. Heskett et al. (2003).
46. Gulati and Oldroyd (2005).
47. Narayandas (2005).
48. Beaujean et al. (2006).
49. Beaujean et al. (2006).
50. Reichhold and Shefter (2000).
51. Dorf, Peppers, and Rogers (2002).
52. Reichheld and Markey (2011).
53. Reichhold and Shefter (2000).
54. Tjemkes et al. (2012).

Chapter 5

1. Cather, Morris, and Wilkinson (2001).
2. Chang (2014).
3. Raynor and Ahmed (2013).
4. Lanen et al. (2013); Kimmel et al. (2012); Kotler and Keller (2011).

References

Akira, I. (2009). *Risk and crisis management 101 cases* (Rev. ed.). Singapore: World Scientific Publishing.

Atkinson, A. A., Kaplan, R. S., Matsumura, E. M., & Young, S. M. (2011), *Management Accounting: Information for Decision-Making and Strategy Execution*, (6th ed.) Chapter 5: Activity Based Cost Systems, Upper Saddle River, NJ: Pearson Education.

Babson, R. W. (2005). *Business fundamentals: How to become a successful business man*. New York, NY: Cosimo Classics.

Baker, H. K., & English, P. (2011).*Capital budgeting valuation: Financial analysis for today's investment projects*. Hoboken, NJ: John Wiley.

Bamber, L. S., & Hughes, K. E. II (2001). Activity based costing in the service sector—The Buckeye National Bank. *Issues in Accounting Education, 16*(3), 381–408.

Barnett, F. W. (2009). *Four steps to forecast total market demand*. Download PDF. Boston, MA: Harvard Business School Press.

Baron, S., Cassidy, K., Harris, K., & Hilton, T. (2009). *Service marketing: Text and cases* (3rd ed.). New York, NY: Palgrave MacMillan.

Beaujean, M., Davidson, J., & Madge, S. (2006). The "moment of truth" in customer service. *McKinsey Quarterly, 1*, 62–73.

Bedbury, S., & Fenichell, S. (2003). *A new brand world: 8 principles for achieving brand leadership in the 21st century*. New York, NY: Penguin.

Berman, K., Knight, J., & Case, J. (2013). *Financial intelligence for entrepreneurs: A manager's guide. Skonieczny to knowing what the numbers mean* (Rev. ed.). Boston, MA: Harvard Business Review Press.

Berry, L. L., & Seltman, K. D. (2007). Building a strong services brand: Lessons from Mayo Clinic. *Business Horizon, 50*, 199–209.

Black, F., & Scholes, M. (1973). The pricing of options and corporate liabilities. *Journal of Political Economy, 81*(3), 637–654.

Bragg, S. M. (2002). *Accounting reference desktop*. New York, NY: John Wiley.

Bragg, S. M. (2012). *Business ratios and formulas: A comprehensive guide* (3rd ed.). Hoboken, NJ: John Wiley.

Brealey, R. A., Myers, S. C., & Alen, F. (2013). *Principles of corporate finance* (11th ed.). New York, NY: McGraw-Hill Higher Education.

Brigham, E. F., & Ehrhardt, M. C. (2013). *Financial management: Theory and practice* (14th ed.). Boston, MA: South-Western College Publishing.

Briner, R. F., Alford, M., & Noble, J. A. (2003). Activity based costing for state and local governments. *Management Accounting Quarterly*, *4*(3), 8–14.

Brown, R. (2010). *Business essentials for utility engineers.* Boca Raton, FL: CRC Press.

Buttross, T., & Schmelzle, G. (2008). Activity based costing in the public sector. In E. M. Bergman & J. Rabin (Eds.), *Encyclopedia of public administration and public policy* (2nd ed.). Boca Raton, FL: CRC Press.

Canada, J. R., Sullivan, W. G., Kulonda, D. J., & White, J. A. (2004). *Capital investment analysis for engineering and management* (3rd ed.). Upper Saddle River, NJ: Prentice Hall.

Cather, H., Morris, R., & Wilkinson, J. (2001). *Business skills for engineers and technologists.* Boston, MA: Newnes.

Catucci, B. (2002). *Ten lessons for implementing the balanced scorecard.* Harvard Business School Balanced Scorecard Report #B0301E. Boston, MA: Harvard Business School Press.

Chaffey, D., Ellis-Chadwick, F., Johnston, K., & Mayer, R. (2009). *Internet marketing: Strategy, implementation and practice* (4th ed.). Upper Saddle River, NJ: Prentice Hall.

Chang, C. M. (2014). *Achieving service success: Maximizing enterprise performance through innovation and technology.* New York, NY: Business Expert Press.

Charnes, J. (2012). *Financial modeling with crystal ball and excel plus website* (2nd ed.). Hoboken, NJ: John Wiley.

Chemey, A., & Kotler, P. (2012). *Strategic marketing management* (7th ed.). Chicago, IL: Cerebellum Press.

Cokins, G. (2006). *Activity-based cost management in government* (2nd ed.). Tysons Corner, VA: Management Concepts.

Coplin, B. (2012). *10 things employers want you to learn in college: The skills you need to succeed* (Rev. ed.). New York, NY: Ten Speed Press.

Cox, L. A. (2010). *Risk analysis of complex and uncertain systems.* New York, NY: Springer.

Delong, T. J., Brackin, W., Cabanas, A., Shellhammer, P., & David, L.(2005). *Proctor & Gamble: Global business services.* Harvard Business School Case #9-404-124. Boston, MA: Harvard Business Review Press.

Dev, C. S., Schulze, H. G., Keller, J., Lane, K., & Frampton, J. (2008). The corporate brand, help or hindrance. *Harvard Business Review*, *86*(2), 49–58.

Dolan, R. J., & Gourville, J. T. (2005, September 22). *Principles of pricing.* Harvard Business School Note #506021. Boston, MA: Harvard Business School Press.

Dorf, B., Peppers, D., & Rogers, M. (2002). *Is your company ready for one-to-one marketing?* Harvard Business School OnPoint Article # 8954. Boston, MA: Harvard Business School Press.

Droms, W. G. (2003). *Finance and accounting for nonfinancial managers: All the basics you need to know* (5th ed.). New York, NY: Basic Books.

Edvardsson, B., Gustafsson, A., Kristensson, P., & Magnusson, P. (2006). *Involving customers in the new service development.* London, UK: Imperial College Press.

Ehrlich, E., & Fanelli, D. (2012). *The financial services marketing handbook: Tactics and techniques that produce results* (2nd ed.). New York, NY: Bloomberg Press.

Ennew, C., & Waite, N. (2013). *Financial services marketing: An international guide to principles and practices* (2nd ed.). New York, NY: Routledge.

Ferrell, O. C., & Hartline, M. (2010). *Marketing strategy* (5th ed.). Independence, KY: Cengage Learning.

Ferri, F., Ferris, W. P., Treadwell, S., & Desai, M. A. (2006). *Understanding economic value added* (Rev. ed.). Harvard Business School Note #206016. Boston, MA: Harvard Business School Press.

Fichman, R. G., & Kemerer, C. F. (2002). Activity base costing for component-based software development. *Information Technology and Management,3*, 137–160.

Fields, E. (2011). *The essentials of finance and accounting for nonfinancial managers* (2nd ed.). New York, NY: AMACOM.

Fitzpatrick, H. (2013). *Marketing management for non-marketing managers.* New York, NY: AICPA.

French, T. D., & Knudsen, T. R. (2007). Marketing in transition. *McKinsey Quarterly*, (3), 5.

Frenkel, I. B., Karagrigoriou, A., Lisnianski, A., & Kleyner, A. V. (2013). *Applied reliability engineering and risk analysis: Probabilistic models and statistical inference*. Hoboken, NJ: John Wiley.

Glass, B. (2012). *Great legal marketing: How smart lawyers think, behave and market to get more clients, make more money, and still get home in time for dinner.* New York, NY: Glazer Kennedy Pub/Morgan James Publishing.

Gourville, J., & Soman, D. (2002). Pricing and the psychology of consumption. *Harvard Business Review*, *80*(9), 90–96.

Greimer, D., Zeithaml, V. A., & Bitner, M. J. (2012). *Service marketing* (6th ed.). Boston, MA: McGraw-Hill/Irwin.

Grenci, R. T., & Watts, C. A. (2007). *Maximizing customer value via mass-customized e-consumer services.* Business Horizon Article # BH226. Bloomington, IN: Indiana University Press.

Griffin, J. (2002). *Customer loyalty: How to earn it, how to keep it* (Rev. sub. ed.). San Francisco, CA: Jossey-Bass.

Gulati, R., & Oldroyd, J. B. (2005). The quest for customer focus. *Harvard Business Review*, *83*(4), 92–101.

Hartley, R., & Claycomb, C. (2013). *Marketing mistakes and successes* (12th ed.). New York, NY: John Wiley.

Healy, P. M. (2001). *Revenue recognition.* Harvard Business School Note, No. 9-101-017. Boston, MA: Harvard Business School Press.

Healy, P. M., & Choudhary, P. (2001A). *Asset reporting.* Harvard Business School Note, No. 9-101-014. Boston, MA: Harvard Business School Press.

Healy, P. M., & Choudhary, P. (2001B). *Liabilities reporting.* Harvard Business School Note, No. 9-101-016. Boston, MA: Harvard Business School Press.

Healy, P. M., & Choudhary, P. (2001C). *Expenses recognition.* Harvard Business School Note, No. 9-101-015. Boston, MA: Harvard Business School Press.

Healy, P. M., & Cohen, P. (2000). *Financial statement and ratio analysis.* Harvard Business School Note, No. 9-101-029. Boston, MA: Harvard Business School Press.

Heskett, J. L., Sasser, W. E., & Schlessinger, L. A. (2003). *The value profit chain: Treat employees like customers and customers like employees.* New York, NY: Free Press.

Hoffman, D., & Bateson, J. E. G. (2010). *Services marketing: Concepts, strategies and cases* (4th ed.). Independence, KY: Cengage Learning.

Horngren, C. T., Foster, G., Datar, S. M., & Rajan, M. T. (2011). *Cost accounting: A managerial emphasis* (14th ed.). Upper Saddle River, NJ: Prentice Hall.

Ittelson, T. (2009). *Financial statements: A step-by-step guide to understanding and creating financial reports.* (Rev. ed.). Pompton Plains, NJ: Career Press.

Jones, R. F. (2006). *Power marketing of architectural services: A critical look at the services provided by architects and designers.* Bloomington, IN: Trafford Publishing.

Kaplan, R. S., & Norton, D.(2007). *Using the balanced scorecard as a strategic management system,* HBR Classic # R0707M. Boston, MA: Harvard Business Review Press.

Keegan, W. J. (2013). *Global marketing management* (8th ed.). Upper Saddle River, NJ: Prentice Hall.

Kelleher J. C., & MacCormack, J. J. (2005). Internal rate of return: A cautionary tale. *McKinsey Quarterly* (Special ed.), 70–75.

Keller, K. L. (2000). The brand report card. *Harvard Business Review, 78*(1), 147–157.

Keller, K. L., Sternthal, B., & Tybout, A. (2002). Three questions you need to ask about your brand. *Harvard Business Review, 80*(9), 80–86.

Kenny, D., & Marshall, J. E. (2000). Contextual marketing: The real business of the internet. *Harvard Business Review, 78*(6), 119–125.

Kimmel, P. D., Weygandt, J. J., & Kieso, D. E. (2012). *Financial accounting: Tools for business decision making* (7th ed.). Hoboken, NJ: John Wiley.

Koller, T., Goedhard, M., & Wessels, D. (2010).*Valuation: Measurement and managing the value of companies* (5th ed.). Hoboken, NJ: John Wiley.

Kotler, P. R., Bowen, J. T., & Makens, J. (2013). *Marketing for hospitality and tourism,* (6th ed.). Upper Saddle River, NJ: Prentice Hall.

Kotler, P., & Keller, K. (2011). *Marketing management* (14th ed.). Upper Saddle River, NJ: Prentice Hall.

Lacey, N.J., & Chambers, D. R. (2010). *Modern corporate finance: Theory and practice* (6th ed.). Plymouth, MI: Hayden-McNeil Publishing.

Lanen, W., Anderson, S., & Maher, M. (2013). *Fundamentals of cost accounting* (4th ed.). Boston, MA: McGraw-Hill/Irwin.

Lovelock, C., & Wirtz, J. (2010). *Service marketing* (7th ed.). Upper Saddle River, NJ: Prentice Hall.

Luehrman, T. A. (1997). What's it worth? A general manager's guide to valuation. *Harvard Business Review, 75*(3), 132–142.

Luerhman, T. A. (2009). *Real options exercise* (Rev. ed.). Harvard Business School Exercise #208045. Boston, MA: Harvard Business School Press.

Makoujv, R. (2010). *How to read a balance sheet: The bottom line of what you need to know about cash flow, assets, debt, equity, profit and how it all comes together.* New York, NY: McGraw-Hill.

Marks, D., Hernandez-Blades, C., Ferguson, R., Tsukahara, C., Volturo, J., Minnec, J., Fisher, M., Gavales, L., Thomas, S., & Hope, R. (2012). *Lessons from the top: Leading CMOs share their case studies.* San Rafael, CA: ExecSense.

McCubbrey, D. (2001). *Business fundamentals.* Seattle, WA: CreateSpace, an Amazon.com company.

Miller P., & Palmer, R. (2000). *Nuts, bolts and magnetrons: A practical guide for industrial marketers.* New York, NY: John Wiley.

Moon, Y. (2005). *Break free from the product life cycle*. Boston, MA: Harvard Business School Note #Ro5o5E. Boston: Harvard Business School Press.

Moore, K. R. (2012). *Using activity-based costing to improve performance: A case study report.* Charleston, SC: BiblioScholar, an imprint of Biblio Bazaar.

Mun, J. (2010). *Modeling risks: Applying Monte Carlo simulation, strategic real options, stochastic forecasting, and portfolio optimization* (2nd ed.). Hoboken, NJ: John Wiley.

Narayandas, D. (2005). Building loyalty in business markets. *Harvard Business Review, 83*(9), 131–139.

Neufeld, D. (2005). *RBC investments: Portfolio planning initiative* (Rev. ed.). Ivey School of Business Case #905E05. Ontario, Canada: University of Western Ontario Press.

Nguyen, H. T., & Walker, E. A. (2005). *A first course in fuzzy logic* (3rd ed.). Boca Raton, FL: Chapman and Hall/CRC.

Nimocks, S. P. (2005). Managing overhead costs. *McKinsey Quarterly*, (2), 106–117.

Nunes, P. F., & Merrihue, J. (2007). The continuing power of mass advertising. *MIT Sloan Management Review, 48*(2), 63–69.

Palmer, A. (2011). *Principles of services marketing* (6th ed.). New York, NY: McGraw-Hill Higher Education.

Rapier, D. M. (1996). *Standard costs and variance analysis.* Harvard Business School Note, No. 9-196-121. Boston, MA: Harvard Business School Press.

Rappaport, A. (2006). 10 Ways to create shareholder value. *Harvard Business Review, 84*(9), 66 –77.

Raynor, M. E., & Ahmed, M. (2013). *The three rules: How exceptional companies think.* New York, NY: Portfolio Hardcover, an imprint of Penguin Group.

Reeve, J., Warren, C. S., & Duchac, J. (2011). *Financial and managerial accounting using excel for success.* Cincinnati, OH: South-Western College Pub.

Reichhold, F. F., & Shefter, P. (2000). E-loyalty: Your secret weapon on the web. *Harvard Business Review, 78*(4), 105–113.

Reichhold, F. F., & Markey, R. (2011). *The ultimate question 2.0: How net promoter companies thrive in a customer-driven world* (Rev. ed.). Boston, MA: Harvard Business School Press.

Revsine, L., Collins, D., Johnson, B., & Mittelstaedt, F. (2011). *Financial reporting and analysis* (5th ed.). Boston, MA: McGraw-Hill/Irwin.

Reznik, L., & Dimitroy, V. (Eds.). (2013). *Fuzzy systems design: Social and engineering applications.* New York, NY: Physica/Springer.

Roberts, M. L., & Zahay, D. (2012). *Internet marketing: Integrate online and offline strategies* (3rd ed.). Independence, KY: Cengage Learning.

Ross, S., Westerfield, R., & Jordan, B. (2012). *Fundamentals of corporate finance, alternate value* (10th ed.). Boston, MA: McGraw-Hill/Irwin.

Ross, T. (2004). *Fuzzy logic with engineering applications* (2nd ed.). Hoboken, NJ: John Wiley.

Rust, R. T., Zeithaml, V. A., & Lemon, K. N. (2004). Customer-centered brand management. *Harvard Business Review, 82*(9), 110–118.

Sanford, R. A. (2011). *The impact of activity-based costing on organizational performance.* UMI Dissertation Publication, Ann Arbor, MI: ProQuest.

Schultz, M., Doerr, J. E., & Federiksen, L. (2013). *Professional services marketing: How the best firms build premier brands. Thriving lead generation engines, and cultures of business development success* (2nd ed.). Hoboken, NJ: John Wiley.

Skonieczny, M. (2012). *The basic of understanding financial statements: Learn how to read financial statements by understanding the balance sheet, the income statement, and the cash flow statement.* Waldorf, MD: Investment Publishing.

Smith, G. N. (2011). *Excel applications for accounting principles* (4th ed.). Independence, KY: Cengage Learning.

Smith, M., & Kawasaki, G. (2011). *The new relationship marketing: How to build a large, loyal, profitable network using the social web.* Hoboken, NJ: John Wiley.

Smock, D. A. (2003, February 6). Ten commandments of reverse auctions. *Purchasing, 132*(2).

Standard & Poor's. (2012). *Standard & Poor's 500 guide 2013.* New York, NY: McGraw-Hill.

Stark, J. (2011). *Product lifecycle management: 21st century paradigm for product realization* (2nd ed.). New York, NY: Springer.

Strouse, K. (2004). *Customer-centered: Telecommunications services marketing.* Norwood, MA: ArTech House Publishers.

Subramanyam, K. R., & Wild, J. (2013). *Financial statement analysis* (11th ed.). Boston, MA: McGraw-Hill/Irwin.

Sullivan, W., Wicks, E. M., & Koelling, C. P. (2011).*Engineering economy* (15th ed.). Upper Saddle River, NJ: Prentice Hall.

Swenson, D., Ansari, S., Bell, J., & Kim, I. W. (2003). Best practices in target costing. *Management Accounting Quarterly, 4*(2), 12–17.

Sydsaeter, K., & Hammond, P. (2012). *Essential mathematics for economic analysis* (4th ed.). Upper Saddle River, NJ: Prentice Hall.

Thomas, R. K. (2009). *Health sciences marketing*. Chicago, IL: Health Administration Press.

Titman, S., John, D., & Martin, J. D. (2011). *Valuation: The art and science of corporate investment decisions* (2nd ed.). Upper Saddle River, NJ: Prentice Hall.

Tjemkes, B., Vos, P., & Burgers, K. (2012). *Strategic alliance management*. New York, NY: Routledge.

Touran, A. (2003). Calculation of contingency in construction projects. *IEEE Transactions on Engineering Management, 50*(2), 135–140.

Tregoe, B. B., & Kepner, C. H. (2006). *The rational manager* (paperback). Princeton, NJ: Princeton Research Press.

Troy, L. (2012). *Almanac of business and industrial financial ratios 2013* (44th ed.). Riverwoods, IL: CCH, Inc.

Vishwanath, V., & Mark, J. (1997). Your brand's best strategy. Harvard *Business Review, 75*(3), 123–129.

Walker, Russell (2013). *Winning with Risk Management*, Singapore: World Scientific Publishing.

Wang, H. (2012). *Monte Carlo simulation with applications to finance*. Boca Baton, FL: Chapman and Hall/CRC.

Ward, S., Light, L., & Goldstine, J. (1999). What high-tech managers need to know about brands. *Harvard Business Review, 77*(4), 85–95.

Ward, S., Light, L., & Goldstine, J. (1999). What high-tech managers need to know about brands. *Harvard Business Review, 77*(4), 85–95.

Waters, H., Abdalla, H., Santillan, D., & Richardson, P. (2003). *Application of activity based costing in a Peruvian NGO healthcare system. Operations Research Results, 1*(3). Revised, published for the U.S. Agency for International Development (USAID) by the Quality Assurance Project (QAP), Bethesda, Maryland.

Waugh, T. (2004). *101 Marketing strategies for accounting, law, consulting and professional services firms*. Hoboken, NJ: John Wiley.

Wheeler, A. (2012). *Designing brand identity: An essential guide for the whole branding team* (4th ed.). Hoboken, NJ: John Wiley.

Wiese, N. (2013). *Activity-based costing (ABC)*. Munich, Germany: GRIN Verlag.

Wirtz, J., Lovelock, C., & Chew, P. (2012). *Essentials of services marketing* (2nd ed.). Upper Saddle River, NJ: Pearson Education.

Yankelovich, D., & Meer, D. (2006). Rediscovering market segmentation. *Harvard Business Review, 84*(2), 122–131.

Index

A
ABC. *See* Activity-based costing
Accounting terms, 8–10
Accounts payable, 84
Accounts receivable, 83
Accumulated depreciation, 84
Acquisition and joint ventures, 126
Activity, 99–100
Activity-based costing (ABC), xi, 7, 11–12,16–29
  in banking and financial services, 21–26
  in governments, 28–29
  in health care, 26–27
  in manufacturing, 16–21
  in software development, 29
  steps to implement, 13–15
  tips for, 15–16
  uses in companies, 12–13
Aeroflot, 201
Ahmed, M., 222
Akira, I., 33
Assess financial feasibility, 219
Asset turnover ratio, 99
Assets, 81, 91–92
Atkinson, A. A., 11

B
Babson, R. W., 5
Balance sheet, 81–88
Balanced scorecard, 115–117
Barnett, F. W., 151
Bell, Alexander Graham, 6
Berry, L. L., 170
Beta distribution function, 39
Better before cheaper, 222
Black-Scholes option-pricing model (BSOPM), 124
Block-flow diagram, 13, 14
Bonds, 85
Book value, 85, 94
Brands (products//services), 166–176
  Dead-end, 170, 171
  High-road, 169, 171
  Hitchhiker, 170, 171
  Low-road, 170, 171
Briner, R. F., 29
Brown, R., 5
BSOPM. *See* Black-Scholes option-pricing model
Buckeye National Bank, 21–25
Budget, 9
Buffalo Best, 112–113
Business fundamentals, 1–2

C
Canada, J. R., 37
Capital assets valuation, 117–132
  acquisition and joint ventures, 126–131
  discount cash flow, 118–119
  internal rate of return, 119–120
  Monte Carlo simulations, 122
  multipliers, 120–122

opportunities-real options, 122–126
Capital surplus, 85
Capitalization, 100–101
Cash, 82
Cash flow, 79
Cather, H., 5, 217
Catucci, B., 117
Certified management accountants (CMAs), 75
Certified public accountants (CPAs), 75
CGS. *See* Cost of goods sold
Channels—communications
marketer-controlled, 187
consumer-controlled, 187
Chekhow, Anton, 218
Close-enough data, 16
CMAs. *See* Certified management accountants
Coca-Cola, 12
Collection period ratio, 99
Competitor focus, 143
Comparison of alternatives, 49
Compound interest, 61
Compound interest equations, 60–66
Constant dollar, 61
Consumer price index, 61
Contingency cost estimation, 43–44
Contribution margin, 9
Convertible bonds, 85
Coplin, B., 2
Cost accounting and control
accounting terms, 8–10
activity-based costing (ABC), 16–29
comparison of alternatives, 49
compound interest equations, 60–66
cost analysis, 59–60
cumulative distribution function, 72–74
depreciation accounting, 66–69
inventory accounting, 69–72
overhead costs, reduction of, 48
probability density function, 72–74
product and service costing, 10–16
project evaluation criteria, 54–58
risk analysis and cost estimation, 33–47
simple cost-based decision models, 49–53
target costing, 29–32
terms, 8–10
time value of money, 60–66
Cost analysis, 59–60
Cost center, 8
Cost drivers, 9, 13, 32
Cost estimation, 33–47
Cost estimation by simulation, 34
Cost model, 34, 40–41
Cost objects, 9, 10, 13
Cost of capital, 94
Cost of goods sold (CGS), 7, 16, 78, 81
Cost pool, 9
CPAs. *See* Certified public accountants
Cumulative distribution function, 36, 42, 72–74
Crystal Ball risk analysis software, 35
Current assets, 82
Current costs, 9
Current liability, 84
Current ratio, 99
Customer feedback, 208–209
Customer focus, 143, 203
Customer interactions and loyalty, 206–211
Customer service, moments of truth in, 205–206

D

Debentures, 85
Debit to equity ratio, 101
Decision trees, 44, 45

Deferred income, 84
Deferred income tax, 85
Depreciation, 79
Depreciation accounting, 66–69
Depreciation accounting base, 94
Direct costs, 8
Direct labor costs (DL), 16
Direct material costs (DM), 16
Discounted cash flow (DCF), xi, 76
Distributing channels, 195
Distribution, types of, 195–196
Diversification, 44
Dividend, 79
Dividend payout ratio, 101
Drucker, Peter, vii, 141

E

E-commerce, on distribution, 197
Earning before interest and taxes (EBIT), 79
Earning per share (EPS), 79, 101
eBay, 207
EBIT. *See* Earning before interest and taxes
Economic value added (EVA), xi, 107–109, 115
Effective interest rate, 61
EPS. *See* Earnings per share
EVA. *See* Economic value added
Exclusive distribution, 196
Expected value, 138
Expenses, 79, 93

F

Factory overhead costs (FO), 16
FASB. *See* Financial Accounting Standards Board
Fichman, R. G., 31
FIFO. *See* First in and first out
Finance, managerial, 75
Financial accounting and analysis
  balanced scorecard, 115–117
  capital assets valuation, 117–132
  financial analysis, fundamentals of, 98–115
  key financial statements, 78–97
  principles, 76–78
  risks, 137–140
  T-accounts, 134–137
Financial accounting principles
  accrual principle, 77
  conservatism, 77–78
  dual aspects, 77
  full disclosure principle, 77
  going concern, 78
  matching, 77
Financial Accounting Standards Board (FASB), 93
Financial analysis, fundamentals of, 98–115
Financial statements
  balance sheet, 81–87
  caution in reading, 93–97
  funds flow statement, 87–90
  income statement, 78–81
  linkage between statements, 90
First in and first out (FIFO), 69
Fixed assets, 84
Fixed costs, 8
Ford Foundation, 7
Franklin, Benjamin, 59
Funds flow statement, 87–90
Fuzzy logic systems, 44

G

GAAP. *See* Generally Accepted Accounting Principles
Generally Accepted Accounting Principles (GAAP), 8, 75
Goldstine, J., 167, 190
Granularity of ABC Model, 15
Gross margin, 79
Gross margin to sales ratio, 100
Gulati, R., 204

H

Honeywell Inc., 12

I

Income statement, 78–81
Indirect costs, 8
Intensive distribution, 196
Inter-functional coordination, 143
Interest, 61
Interest coverage ratio, 100
Internal rate of return (IRR), xi, 55–56
Inventory, 83
Inventory accounting, 69–72
Inventory accounting method, 94
Inventory costs, 8
Inventory turnover ratio, 99
IRR. *See* Internal rate of return

J

Jordan, David Starr, 224

K

Kaplan, R. S., 116, 117
Keller, K. L., 168
Kemerer, C. F., 31
Kenny, D., 191
Kepner-Tregoe method, 111
Knowledge economy, 176

L

Last in and first out (LIFO), 69
Lean Six Sigma, 4
Liabilities, 84, 92
LIFO. *See* Last in and first out
Light, L., 190
Liquidity, 98–99
Long-term debt to capitalization ratio, 100
Long-term liabilities reportable, 94
Long-term liability, 85
Lovelock, C., 161, 162
Loyalty, 206–211
Luehrman, T. A., 118

M

Market segmentation
  criteria for, 154
  pitfalls of, 154–155
  purpose of, 153
  rediscovery of, 155–158
  steps in, 153–154
Market to book ratio, 101
Market values, 94
Marketers, 147
  customers, 148
  environment, 148
  market, 148, 151
  segment drivers and model, 151
  sensitivity analysis, 151–152
Marketing
  contextual, 191–192
  forecast, 150–152
  mix (seven Ps), 158–202
  process, 143–144
  segmentation, 152–158
  service value *vs.* knowledge/technology, 149
  word-of-mouth campaign, 172
Marketing effectiveness diagram, 145
Marketing function
  key elements in, 147–150
  marketing process, 143–145
  product and service marketing, 146–147
  sales *vs.* marketing, 142–143
Marketing interaction, 142
Marketing management
  alliances and partnerships, 211
  consumer survey, 214–215
  customers, 203–211
  function of, 142–150
  market forecast, 150–152
  market research, 214–215
  market segmentation, 152–158
  organizational effectiveness, 211–213
  people, 200–202
  physical evidence, 200
  placement (distribution) strategy, 195–200
  pricing strategy, 176–186

process design, 200
product (service) strategy, 159–176
promotion (marketing communications), 186–194
Marketing orientation, 143
Marketing process, 144
Marshall, J. E., 191
MaxSalud Institute for High Quality Healthcare, 27
Mayo Clinic, 172
McCubbrey, D., 5
McKinsey & Company, 201
Meer, D., 155
Moments of truth, in customer service, 205–206
Monte-Carlo simulating method, 3, 7, 35, 122, 151
Multiple-period analyses, 60

N

Narayandas, D., 203
Net fixed assets, 84
Net income, 79, 95
to OE ratio, 100
to sales ratio, 100
to total asset ratio, 100
Net present value (NPV), 3, 54–55
Nominal dollar, 61
Nominal interest rate, 61
Nonactivity costs, 11
Normal probability density function, 38
Norton, D., 116, 117
NPV. *See* Net present value

O

Oldroyd, J. B., 204
Operations—Assets in place, 118–122
Opportunity costs, 9–10
Opportunities—real options, 122–126
Option, 123
Organizational effectiveness, 211–212
Organizational structures, 196–197
Other assets, 84
Overhead costs, reduction of, 48
Owners' equity, 85

P

P/E. *See* Price to earning ratio
Payback period (PB), 56
PB. *See* Payback period
Performance ratios, 98–103
Peters, Tom, 48
Physical evidence, 200
PI. *See* Profitability index
Placement (distribution strategy), 195–200
Poisson distribution functions, 39
Prepaid expenses, 83
Price to earning ratio (P/E), 101
Price-quality relationship, 178
Pricing strategy
factors affecting price, 177–181
penetration strategies, 176–177
pricing and psychology of consumption, 185–186
pricing methods, 181–185
skimming strategies, 176–177
value-added pricing, 183
Probability density function, 42, 72–74
Process design, 200
Product (service) strategy
brands, 166–176
composition, 161–163
industrial *vs.* consumer products/services, 160
life cycles, 164–165
portfolio, 165–166
positioning, 163–164
Product and service costing, 10–16
customer experience, 146–147
lifecycle, 146
time and place of production, 146–147
variability in, 147
Products/services innovations, 221

Profitability, 100
Profitability index (PI), 56–57
Profit orientation, 144
Project evaluation criteria, 54–58
Promotion (marketing communications)
  communication process, 187–188
  contextual marketing, 191–192
  high-tech and consumer products/services, 189–190
  internet-enabled communication options, 190–191
  marketing to segment of one, 193
  mass advertising, 193
  middle of road approach, 193–194
  promotion strategy, 189

Q

Quick ratio, 99

R

R&D, 81
Ratio analysis, 104–107
Raynor, M. E., 222
Reichhold, F. F., 208
Relative position of power, 178–185
Replacement evaluation, 52
Retained earnings, 85
Return on invested capital, 100
Return on investment (ROI), 98
Revenue, 92–93
Revenue before cost, 222
Risks, 137–140
  Mathematical representation, 33–34
  Management, 140
  Risk pooling, 36–37
Risk analysis and cost estimation, 33–47
Risk analysis software, 35
ROI. *See* Return on investment
Rust, R. T., 172

S

Sales, 80–81
Sales general and administrative (SG&A), 96
Science, technology, engineering and math (STEM), 1, 2
Sales revenue, 78
Sales to employee ratio, 100
Selective distribution, 196
Self-study, 2, 4
Seltman, K. D., 170
Sensitivity analysis, 43
SG&A. *See* Sales general and administrative
Shefter, P., 208
Simple cost-based decision models, 49–53
Singapore Airlines, 201
Single-period analysis, 59–60
Standard costs, 9
Step function costs, 9
Stock, 85
Stock price, 85
Stock value, 101
Sunk-cost effect, 185
Sunk costs, 10

T

T-accounts, 134–137
Target costing, 29–32
  principles of, 31
Target pricing, 183, 184
Texas Department of Agriculture (TDA), 28
Three-point estimate, 35
Time value of money, 60–66
Triangular probability density function, 38

U

Unit costs, 15, 220
United States Agency for International Development (USAID), 27
United Way, 7

V
Value-added pricing, 183
Value Profit Chain Model, 200
Vanguard Group, 207
Variable costs, 8
Variance, 9
Vision, 217

W
Walker, Russell, 140
Ward, S., 167, 190
Waterfall approach for software
  development, 29
Wiese, N., 10
Wirtz, J., 161, 162
Working capital, 99
Working capital turnover ratio, 99

Y
Yankelovich, D., 155

FORTHCOMING TITLES FROM OUR ENGINEERING MANAGEMENT COLLECTION

Dr. C. M. Chang, Adjunct Professor Emeritus, State University of New York at Buffalo, Editor

Analytics and Tech Mining for Engineering Managers
By Scott W. Cunningham, Jan H. Kwakkel

The Entrepreneurial Engineer
By Anita Leffel, Cory Hallam, William Flannery

Engineers Guide to Early Risk Identification and Assessment with RED
By Katie Grantham

Innovation Management for Engineers
By Rob Dekkers

Six Sigma and Statistical Tools for Engineers and Engineering Managers
By Wei Zhan, Xuru Ding

Momentum Press is looking for authors in this collection and in many more engineering and science areas. For more information on how to join the Momentum Press author team, please visit www.momentumpress.net/collections.

Announcing Digital Content Crafted by Librarians

Momentum Press offers digital content as authoritative treatments of advanced engineering topics, by leaders in their fields. Hosted on ebrary, MP provides practitioners, researchers, faculty and students in engineering, science and industry with innovative electronic content in sensors and controls engineering, advanced energy engineering, manufacturing, and materials science. **Momentum Press offers library-friendly terms:**

- perpetual access for a one-time fee
- no subscriptions or access fees required
- unlimited concurrent usage permitted
- downloadable PDFs provided
- free MARC records included
- free trials

The **Momentum Press** digital library is very affordable, with no obligation to buy in future years.